Linguistic and Genetic (mtDNA) Connections between Native Peoples of Alaska and California

Linguistic and Genetic (mtDNA) Connections between Native Peoples of Alaska and California

Ancient Mariners of the Middle Holocene

Cecil H. Brown
Kent G. Lightfoot
Nancy J. Turner
and
Dana Lepofsky

LEXINGTON BOOKS
Lanham • Boulder • New York • London

Published by Lexington Books
An imprint of The Rowman & Littlefield Publishing Group, Inc.
4501 Forbes Boulevard, Suite 200, Lanham, Maryland 20706
www.rowman.com

86-90 Paul Street, London EC2A 4NE

British Library Cataloguing in Publication information available

Library of Congress Cataloging-in-Publication Data

Names: Brown, Cecil H., 1944- author. | Lightfoot, Kent G., 1953- author. | Turner, Nancy J., 1947- author. | Lepofsky, Dana, 1958- author.
Title: Linguistic and genetic (mtDNA) connections between native peoples of Alaska and California : ancient mariners of the middle Holocene / Cecil H. Brown, Kent G. Lightfoot, Nancy J. Turner, Dana Lepofsky.
Description: Lanham : Lexington Books, 2024. | Includes bibliographical references and index.
Identifiers: LCCN 2024030361 (print) | LCCN 2024030362 (ebook) | ISBN 9781666915105 (hardcover) | ISBN 9781666915112 (epub)
Subjects: LCSH: Aleut language—History. | Penutian languages—History. | Comparative linguistics. | Languages in contact. | Human population genetics—Alaska. | Human population genetics—California.
Classification: LCC PM31 .B76 2024 (print) | LCC PM31 (ebook) | DDC 979.4004/9741—dc23/eng/20240731
LC record available at https://lccn.loc.gov/2024030361
LC ebook record available at https://lccn.loc.gov/2024030362

♾™ The paper used in this publication meets the minimum requirements of American National Standard for Information Sciences—Permanence of Paper for Printed Library Materials, ANSI/NISO Z39.48-1992.

In Memoriam
Knut Bergsland (1914–1998) and Catherine A. Callaghan (1931–2019)

Contents

Acknowledgments

Given the interdisciplinary nature of this book (particularly archaeology, ethnobiology, human molecular genetics, and linguistics), we reached out to many scholars for help and advice during various stages of its preparation. For contributions, we are grateful to Gene Anderson, M. Kat Anderson, Adam St. Arnaud, Brenda Beckwith, Juliette Blevins, Brian Byrd, Lyle Campbell, Roy L. Carlson, Jim Chatters, Brian D. Compton, Douglas Deur, Matthew Dryer, Juan Alvaro Echeverri, Nicholas Q. Emlen, Patience Epps, Daryl W. Fedje, Robert Forkel, Martin Haspelmath, Richard Hebda, John Hewson, Jan Henrik Holst, John R. Johnson, Charles Julian, Frederika Ann Kaestle, Grzegorz Kondrak, Robert Kopperl, Harriet V. Kuhnlein, Clark Spencer Larsen, Ken Lertzman, Bryn Letham, Quentin Mackie, Steven A. Marlett, Duncan McLaren, David E. Mora-Marín, Michael Moratto, Madonna Moss, Christina Munck, Andrey Nikulin, Peter Norquest, Andrew Pawley, Eric Peterson, Donald A. Ringe, Martine Robbeets, Laurent Sagart, Frank Seifart, Fred H. Smith, Patrick G. Saltonstall, George Starostin, Amy F. Steffian, Mark Stoneking, Jan Timbrook, Robert D. Turner, Edward Vajda, Andrea Weiser, and Hein van der Voort. (We apologize for any inadvertent omissions.)

In addition to individuals from whom we sought and received help, we are grateful to several institutions, including the Hakai Institute; the Alutiiq Museum; the Phoebe A. Hearst Museum; the Archaeology Program, National Science Foundation (BCS-1848878); and the Stahl Endowment in the Archaeological Research Facility at U.C. Berkeley.

Thanks are due to editors at Lexington Books, who made the technical production of this work a pleasure: Jana Hodges-Kluck, Madeline Jenkins, Jasper Mislak, and Alexandra Rallo.

Finally, several people deserve acknowledgment for their extraordinary contributions to this book project, each over a number of years providing counsel on specific aspects of the manuscript itself and/or valuable ideas for consideration: David Beck, Anna Berge, Pamela Brown, Catherine A. Callaghan†, Michael D. Fortescue, Eric W. Holman, and Søren Wichmann. Special thanks go to Eric for his quantitative expertise and invaluable insights, without which this book would not have been possible.

Introduction

We describe a genealogical (linguistic) relationship between geographically remote languages spoken by pre-contact Indigenous Peoples of the Pacific coast of North America: Aleut of southwest Alaska and the Utian languages of central California. Historical linguists have not previously recognized these languages as genealogically affiliated. In reviewing evidence for and telling the story of Aleut-Utian, we propose new methodologies for investigating distant relationships, building on the traditional approach of comparative linguistics.

Of importance to all aspects of our story is its timeline, especially that of the Holocene era. For this, we use the geographic chronology reported by Walker et al. (2012):

Early Holocene: 12000–8200 BP
Late Early Holocene: 9000–8200 BP
Middle Holocene: 8200–4200 BP
Late Holocene: 4200–present

Walker et al. (2012) recognize the Pleistocene–Holocene boundary as 11700 BP (BP = years before present).

This is the story. In the Middle Holocene, speakers of the Utian dialect of Proto-Aleut-Utian (PAU), a marine food-harvester group, began a journey from their homeland in coastal southwest Alaska, in the Kodiak Island area, following the North Pacific coastline to the San Francisco Bay Area of central California. These travelers are the "Ancient Mariners" of our book title (with apologies to Samuel Taylor Coleridge), since their journey was likely undertaken by watercraft. They left behind in Alaska speakers of their sister dialect, Aleut. At some point after the beginning of the migration, Aleut and Utian dialects developed into distinct languages, respectively, Early Aleut and Proto-Utian. As Utians navigated along the coast, they made stops, some of them of prolonged duration (months or, perhaps, even years). These layovers involved contact with Native local peoples who spoke languages different from their own, such as Proto-Salish and Proto-Wakashan.

Utians ended their journey in coastal central California, where they settled and continued their maritime way of life. Over time, different groups of Utian speakers began to move into interior regions, following rivers and their tributaries upstream that drained into coastal bays. These migrations resulted in geographic separations, leading to the development of different Utian languages, such as Proto-Miwok and Proto-Costanoan and their respective offspring languages.

Meanwhile, Aleut speakers remained in southwest Alaska but, unlike its sister language in California, Aleut did not give rise to daughter languages, at least not any known to us. For several thousand years, Aleut speakers were in close contact with speakers of Proto-Eskimo and its offspring languages, resulting in considerable borrowing and language convergence. Aleuts ultimately migrated to the Aleutian Islands, where many remain to the present.

For at least two centuries, scholars have used the comparative method of historical linguistics to determine whether languages showing similarities to each other are genealogically related. French and Spanish, for example, are genealogically affiliated since they both developed from Latin (Proto-Romance). While relationships for Latinate languages are often historically known, this is not so for most of the world's related languages. The comparative method provides a means for determining language affiliations for non-recorded languages and is recognized today as the gold standard of historical linguistics.

Another approach, quantitative comparison, is a much more recent development, emerging around the same time as electronic digital computers some 85 years ago and gathering steam only in the past two decades. While not all uses of the quantitative approach involve automation, all have in common the use of quantifiable linguistic data and algorithms to which computer analysis can be applied. In contrast, proponents of the traditional approach rarely if ever quantify data but rely instead on their typically rich personal knowledge of the lexicons and grammars of compared languages. Although not discussed in published literature, the two methodological camps are competitive, with some traditionalists regarding quantifiers as young Turks more interested in computer programs than linguistics, and quantifiers viewing traditionalists as old fogies spending way too much time obtaining meaningful results. In our view, however, the two approaches are complementary, and we combine them to study the Aleut-Utian relationship.

We begin the discussion with a detailed description of our quantitative approach (chapter 1), followed by its application to lexical data from Aleut and Utian languages, and then by the traditional method applied to the same and additional data (chapter 2). (Readers with no or little background in [historical] linguistics may find chapter 1 especially challenging. We suggest these readers first read section 1.1 of chapter 1, skipping then to chapter 2, and returning to chapter 1 after becoming familiar with the book's scope and focus.) After presenting a solid case, in our view, for the kinship of these languages, we proceed to examine other possibly supportive evidence for the relationship, including structural features of the languages (chapter 3), and the use of mitochondrial-DNA analysis (chapter 4). Next, broader language relationships for Aleut and Utian are discussed in detail, including the long-embraced Eskimo-Aleut proposal, which we challenge, and Utian languages as constituents of the California Penutian family (chapter 5).

Chapter 6 examines loanword evidence for the migration of Proto-Utian speakers by watercraft from Alaska to California, accounting for the great geographic removal of historic Aleut and Utian speakers from one another, including reference to local groups encountered along the way speaking languages such as Proto-Wakashan and Proto-Salish. Chapter 7 presents a detailed reconstruction of ecological conditions, many previously unknown to Utian mariners, along the 5000 kilometers of coastline from Alaska to California. Chapter 8 discusses archaeological sites in coastal south Alaska and central California and in between, possibly occupied or visited by Utian speakers in Middle Holocene times. Chapter 8 also makes an important methodological contribution to the archaeology of human migration. Using data compiled in preceding chapters, chapter 9 reconstructs the culture history of Aleut-Utian people, including details relating to homelands, migrations, and interactions with other groups encountered at different times and places. Chapter 10, the conclusion, reviews major contributions of this book, major controversies addressed, and suggestions for continuing study.

The book has six appendices (A–F), each associated with one of six chapters, which can be found online at https://rowman.com/ISBN/9781666915112 beneath the tab labeled *Supplemental Files*.

Chapter 1

Comparative Approaches in Historical Linguistics

1.1 TRADITIONAL COMPARATIVE METHOD

Both traditional and quantitative strategies of historical linguistics focus on language similarity involving systematic relationships such as cognation and sound correspondence. The traditional (or standard) method is essentially the quantitative approach without metrics, while the quantitative approach (at least as used here) is the traditional method with metrics.

Any two or more languages are genealogically related if descended from a common ancestral language (proto-language); for example, French and Spanish are offspring of Latin (Proto-Romance). Linguists use language comparison to determine affiliations among the world's approximately 7,000 languages. Over the past two centuries or so, a single strategy has dominated this endeavor: the comparative method of historical linguistics. The starting point for this approach is to compare languages thought to be possibly related, looking for commonalities in phonology, lexicon, and grammar.

All human languages may be genealogically connected, that is, all descended from a single ancestral language of very remote times (monogenesis). If so, there is little reason to use methods such as the standard and quantitative approaches to show any two or more languages to be related. Most comparative linguists accept that these approaches cannot detect language affiliation among languages separated from one another by more than around 9,000 years. Typically, of greatest interest is finding heretofore unrecognized language relationships close enough to be detected and similar enough to contribute to the reconstruction of the cultural history of peoples who speak them, whether the languages are chronologically removed from one another by 1,000 years or less or by as much as 9,000 years (chapter 9).

A kinship connection for languages is considered detected if observed similarities show regular sound correspondences across the languages compared (Brown et al. 2013). (For a detailed, pedagogical treatment of the traditional method [including exercises] consult Campbell 1999:108–162.) Recognition of sound correspondences is the bedrock of the traditional method of identifying languages descended from a common ancestor. Accurate reconstruction of the phonological inventories and vocabulary of proto-languages are both goals of the standard method (chapter 2). Neither reconstruction is possible, however, in the absence of sound correspondences.

Two sound segments correspond when they occur in cognate words of genealogically related languages (Brown et al. 2013). For example, K'iche' *raʃ* "green" and Huastec *jaʃ* "green" are cognate words since both are descendant forms of the word for "green" in their common ancestral language, Proto-Mayan (Mesoamerica). Corresponding sounds in this example are respectively *r : j, a : a,* and *ʃ : ʃ.* By definition, a sound correspondence is observed when corresponding sounds occur in at least two cognate sets for genealogically related languages. (Sound correspondences typically occur in more than two.) For example, *r : j* is also apparent for K'iche' *war* "sleep"/Huastec *waj* "sleep" and for K'iche' *ra:h* "spicy"/Huastec *jah-* "spicy." Because of such recurrence, these matches are dubbed regular sound correspondences, usually shortened to just sound correspondences.

Sound correspondences are typically products of ways in which the sounds of languages change over time (Brown et al. 2013). For example, the *r : j* correspondence recognized for K'iche' and Huastec is the product of a Proto-Mayan sound segment, **r*, that developed in different ways in Proto-Mayan daughter languages (Campbell 1999:125), being maintained as *r* in K'iche' and shifting to *j* in Huastec. The reconstructed word for "green" in Proto-Mayan is **raʔʃ.* (The raised asterisk [*] on a sound segment or word indicates these to be hypothetical, reconstructed rather than actually observed.)

The comparative method is not always straightforward, however. Language similarities can have origins other than common inheritance from an ancestor. Other sources of resemblance include borrowing, language universals such as onomatopoeia, and just plain coincidence (chance). A problem for comparativists seeking genealogical relationships is how to sort similarities due to kinship from those resulting from other such factors.

Another issue is that the comparative method does not (yet) provide consistency regarding just how much similarity must be observed before languages are officially deemed related. Some comparativists feel comfortable claiming a relationship if they find only a few sound correspondences, while others require more than a few but supply no precise number to be met. Kessler (2001:1) addresses this issue:

> When a linguist has finished compiling lists of correspondences between languages, there remains the problem of judging whether the amount of evidence is strong enough to prove that the languages are . . . connected [. . .] yet there is no commonly agreed upon method for doing so. Linguists looking at the same data can come up with widely differing appraisals as to whether the case for historical connection has been proved or not.

The problem Kessler (2001) raises can be addressed in different ways. For example, he opts for a statistical/probability strategy which, even after more than 20 years, has yet to resolve the issue to the satisfaction of all linguists. In this study, we use an altogether different approach (section 1.2), based on quantitative protocols that combine results of two different comparative procedures, the application of which yields consistent evaluative results. These procedures are "Quantitative Screening," formerly known in a slightly different version as the BWB approach (Brown et al. 2014, Brown 2017, 2020), and ASJP (Automated Similarity Judgment Program) (Brown et al. 2008, Holman et al. 2008, The ASJP Database - (clld.org)).

1.2 QUANTITATIVE SCREENING

The purpose of quantitative screening is to evaluate as objectively as possible findings of the comparative method as presented in theses and dissertations, published books and articles, and personal communications. These sources typically include assembled lists of lexical (word) comparisons for proposed language families. A subset of binary language comparisons is selected for quantitative screening from among lexical comparisons for a proposal of language genealogical relationship. From this subset, weaker comparisons (in terms of semantic similarity) are screened out and the remaining binary comparisons are evaluated, all according to a single protocol (basically a series of iterated algorithms). Quantitative screening supports language connection if this process results in a collection of binary comparisons whose similarity is not likely to be explained by chance; if explained by chance, no language affiliation is likely. Since a full account of the latest version of the quantitative screening procedure has yet to be published, we describe it in its entirety here.

(A recent trend in historical linguistics is the computerization of the comparative method or parts of it [List et al. 2017, List 2019]. While quantitative screening for this study is not digitized, the algorithms used are clearly compatible with automation.)

1.2.1 Standardization

The following requirements for screening impose a measure of uniformity (standardization) on lists of comparative word sets and sound correspondences undergoing quantitative evaluation.

1.2.1.1 Binary Comparisons

Quantitative screening is designed to evaluate only *binary* comparative sets (e.g., K'iche' *raʃ* "green"/Huastec *jaʃ* "green") and, consequently, is not applicable to multilateral comparison (e.g., K'iche' *raʃ* "green"/Huastec *jaʃ* "green"/Mam *ʧaʔʃ* "green" [Kaufman 1970]). The binary approach renders evaluations of comparisons simpler to undertake either manually or through automation than those involving more than two languages. Rigorous methods of automatic multilateral comparison are just beginning to appear (List 2019) and may eventually supplant the current binary comparison procedure. Until then, restriction to binary sets is important since it constitutes an additional strategy to safeguard against the chance development of lexical similarity. Ringe (2002:416) summarizes the essential problem with multilateral comparison as follows:

> the combinatorics of multilateral comparison ensure that as more and more languages are added to the comparison the probability of chance resemblances continues to rise, until it becomes practically impossible to demonstrate that the similarities one has found are in fact greater than would be expected by chance

In other words, as the pool of lexical items from which words similar to a target word increases, the more likely a word similar only by chance to the target item will be found and identified, incorrectly, as a cognate of the target. One way to minimize the possibility of misjudging two coincidentally similar words as being cognates is to limit comparisons to binary matches.

1.2.1.2 Translation Equivalence

A pair of words is included in a list of comparative sets only if their assigned glosses show identity or near identity in meaning. Translation equivalence regularizes the treatment of semantics. The following description is based on Brown (2017).

Translation equivalence is observed when glosses of compared words are identical, for example, *dog/dog,* but not when compared identical words are homophones, for example, *lie* (untruth)/*lie* (prone). Instances of near identity are also accepted, for example, inflected forms of the same stem such as *dog/dogs* and *burn/burned*, and ablaut forms such as *mouse/mice, swim/swam,* and *tear/torn.* Pairs are also included when glosses are not identical but have a "core constituent" in common: for example, *flower/to flower, friend/friendly,* and *acorn woodpecker/woodpecker.* A core constituent is a stem bearing the semantic/functional weight of a complex gloss. For example, *woodpecker* is the core constituent of *acorn woodpecker* because the latter's referent is a woodpecker, not an acorn.

Translation equivalence is also satisfied under other conditions where glosses are not identical: for nomenclaturally unrelated synonyms, for example, *small/little;* or when one gloss of compared words designates a taxonomic type of a "reasonably narrow category" denoted by the gloss of the other word, for example, *spruce/evergreen tree, sapsucker/woodpecker,* and *car/vehicle,* where the second gloss of each match designates the respective category. In contrast, we would not consider categories designated, for example, by *tree* and *bird* to be sufficiently narrow and, consequently, would not accept compared glosses such as *spruce/tree* or *sapsucker/bird.* When two glosses denote members of the same reasonably narrow generic category, they are translation equivalent, for example, *spruce/redwood, sapsucker/acorn woodpecker,* and *car/truck,* where categories are respectively "evergreen tree," "woodpecker," and "vehicle." What may or may not be considered a "reasonably narrow category" may differ from one comparativist to the next, since at present no objectively based definition exists. Consequently, we cannot currently describe quantitative screening as a completely objective approach.

Finally, translation equivalence is considered satisfied when the meanings of two compared words are not strictly the same, but when the meaning of an offspring word of the second comparand is equivalent to the meaning of the first comparand. An example (taken from chapter 2, 2.3 [comparison 16]) is:

A *huluχ-six* "to blink . . . "
PMi **pel-* "pertaining to eyes" (R: Mil *pel-ak* "to blink")

The two compared words in this example are A (Aleut) *huluχ-six* "to blink . . . " (from Bergsland's Aleut dictionary), and PMi (Proto-Miwok) **pel-* "pertaining to eyes" (from Callaghan's Utian dictionary). By the above criteria, translation equivalence does not seem to pertain to this comparison. However, a reflex (R) or offspring word of PMi, that is, Mil (Lake Miwok) *pel-ak* "to blink," shows the same meaning as the Aleut word. We accept this as translation equivalence for the comparison. (Callaghan [2014] chooses "pertaining to eyes" as the reconstructed meaning of the PMi word. However, given the meanings of the reflex word and that of the Aleut word, "to blink" would be more appropriate.)

1.2.1.3 Sound Correspondences

The sole purpose of quantitative screening is to evaluate lexical evidence assembled in support of language affiliation. It is not a substitute for the traditional comparative method since, in addition to attempting to determine relatedness, the traditional method can be used to search for as many sound correspondences as possible to reconstruct the phonological inventory and vocabulary of an ancestral language. Quantitative screening searches out sound correspondences but stops short of going further once the iterative process (section 1.2.1.4) has run its full course and relatedness is determined likely or not.

Several conditions are attended to in determining what is or is not considered a "screened-in" sound correspondence. These are:

Correspondences involving sound segments corresponding with Ø (nil), for example, *k* : Ø, are excluded. The odds of finding a chance sound correspondence are significantly increased when Ø is the corresponding target, because a non-sound "segment" (Ø) can correspond to many sounds, only one of which is usually due to relatedness. Ø can reenter consideration after the iterative stage of evaluation runs its course and is replaced by the full-scale use of the traditional comparative method.

When sound correspondences for a comparative set are limited only to vowels (V : V), the comparative set is screened out. Vowel inventories for languages are usually, if not always, substantially smaller than consonant inventories, and thus vowel correspondences (V : V) are considerably more likely to be found by chance than those with consonants. Requiring at least one consonant correspondence (C : C) per comparative set is an important hedge against coincidental similarity. V : V, like Ø, can reenter consideration after the iterative stage of evaluation runs its course and is replaced by the full-scale use of the traditional comparative method.

When identifying correspondences involving vowels, only *vowel quality* is taken into consideration. Attributes such as vowel nasalization, length, laryngealization, stress, and tone are disregarded, since such features, if marked at all, are often not clearly specified as to function or value in lexical sources. To ensure *consistent treatment* of these features across languages, they are excluded from the screening process, conforming in general with the transcription of words in ASJP (through the use of ASJPcode, see 1.2.3).

When identifying consonant correspondences, consonant length and geminate consonants are not considered. Long consonants are acoustically and orthographically difficult to distinguish from geminate consonants. To avoid misrepresentation in quantitative screening, these are both treated as short, single consonants (C) instead of long (C:) or geminate (CC).

Both consonant length and vowel length (including other attributes of vowels such as nasalization) can reenter consideration when a proposal advances to the level of traditional comparison.

1.2.1.4 Iterative Screening

After standardization, two lists (one for cognates and one for sound correspondences) are produced, which are scrutinized to determine whether there are sufficient sound correspondences to justify each comparative set (at least two sound correspondences required per set) and enough sets to justify each sound correspondence (at least three sets required for quantitative screening). Since the two lists depend on each other, the screening process iterates alternately between them, starting with the cognate list. The iterative process is finished when items from one or the other of the two lists are exhausted.

1.2.1.5 Point Assignment to Lexical Sets

A crucial aspect of quantitative screening is the assignment of points to each binary lexical set assembled in support of a genealogical relationship. The more points found for a comparative set, the more likely the set's similarities are due to a historical connection than to chance. Point assignment is as follows:

Two points for each sound correspondence observed for a comparative set. When correspondences are clearly presented, determining how many correspondences pertain to a set and how many points are accordingly earned is usually straightforward. A comparison receives no points for correspondences, however, if all its correspondences entail *only corresponding vowels*, or if all its correspondences involve *metathesis*, for example, XaY compared to YaX.

One point for a comparison devoid of potential onomatopoeia. If any of the compared words are onomatopoetic, such as words for "dog" (bark), "fire" (crackle), "lightning" (bang), or "to rain" (pitter patter), then no points are awarded for this feature. Sets judged as lacking potential onomatopoeia accordingly earn one point when none of the referents of the words compared entail objects or actions having a characteristic or iconic sound, for example, "moon," "talk," "tree," "eye." The number of points awarded in this category does not depend on whether the forms in question appear to be likely or plausible phonological imitations of the sounds associated with their referents, since this is an inherently subjective process and likely to lead to substantial disagreements among individuals scoring sets.

One point for perfectly matched phonological segments with no unexplained residue. If one word of a comparative set shows a certain number of phonological segments (e.g., ABCD with four) and the other shows the same number (WXYZ), then all segments are matched (with no residue), and the set receives one point. A point is not earned for this feature when one of the compared words of a set shows more segments (ABCD) than the other (WX), and the residue (CD) is not explained, for example, as possibly being some type of affix.

1.2.1.6 Point Cutoff for Sets

Aggregate points for a comparative set are considered indicative of the strength of set convincingness: the more points a set earns, the more likely the similarity of compared words is due to genealogical affiliation rather than to other factors such as chance. A cutoff is therefore defined such that comparisons with enough points to reach it or exceed it are retained while sets below the threshold are screened out. In earlier versions of screening (BWB) described in Brown, Wichmann, and Beck (2014) and Brown (2017, 2020), translation equivalence received two points and the cutoff was set at seven points. Translation equivalence is now required for all sets as part of standardization (section 1.2.1.2), and the equivalent cutoff is five points. This benchmark ensures at least two sound correspondences (and four points) per acceptable set. A set with fewer than two sound correspondences cannot receive more than four points.

1.2.1.7 Cutoff for Number of Correspondences

The list of possible sound correspondences for a collection of comparative sets is inspected to determine how many screened-in sets (each with at least five points) support them. A cutoff of *three* screened-in lexical sets is used. Two sound segments, one each from the two compared languages, must correspond in at least three lexical sets for the two sounds to be screened in as sound correspondences. Correspondences supported by fewer sets are screened out since correspondences supported by only two sets are more likely to occur by chance than those supported by three or more sets.

Sound correspondences do not contribute points to the comparative sets to which they belong if they are screened out, and the iterative process returns to a recount of points for the lexical sets. Sound correspondences are recounted as more sets are screened out, and so on until no more sets or correspondences can be eliminated.

1.2.2 Borrowing and Basic Vocabulary

The major goal of earlier versions of quantitative screening was to recognize comparative sets showing similarities due to kinship (and, by default, to distinguish these from sets with resemblances due to factors such as borrowing and chance). In the current version, the goal is extended to finding not only comparative sets whose similarities are due to kinship, but also those due to borrowing. The traditional comparative method provides no systematic way to recognize similarities attributable to borrowing or loaning words between languages.

Some loanwords are obvious, such as French *weekend* and *shopping* borrowed from English, but often they are not so obvious. Features of loans, such as patterns of distribution across languages and phonological adaptation of a loan to the sound system of the language into which it is borrowed, can render some non-obvious borrowed words apparent, but no algorithmic-quantitative means for doing so is known to us, including the system proposed here. While our quantitative approach to borrowing cannot identify individual words as loans, it does provide a systematic means for determining percentages of groups of words as having been borrowed.

The key to this is basic vocabulary. Words of basic vocabulary designate objects, concepts, and actions universal to human experience and are found in lexicons of all spoken languages, for example, words for "one," "louse," "knee," "sleep," "moon," and "water." Mainstream comparativists have long assumed basic vocabulary to be more resistant to diffusion and, hence, to lexical replacement through borrowing than non-basic vocabulary (Campbell and Poser 2008: 24, 166–167, 174, 194). Brown (2017) provides substantial evidence gathered from Haspelmath and Tadmor (2009a, b) validating this assumption.

Haspelmath and Tadmor (2009a, b) assemble data relating to lexical borrowing from forty-one globally distributed languages. This endeavor entails identifying words in each of these languages for many meanings (1,460 in 2009a and more in 2009b) and having language specialists specify those that are loanwords into these languages and, by default, those that are not. To quantify this information, a borrowing index is calculated for each meaning, which is essentially the number of languages with loanwords for a specific meaning divided by the total number of languages for which a word for the meaning has been reported.

In the appendix to Brown (2017), the use of this information is described in detail to determine whether basic vocabulary (specifically, Swadesh's [1955a] list of 100 meanings) is more resistant to borrowing than is cultural vocabulary. The unequivocal finding is that basic vocabulary is, on average, around *four times less likely* to be borrowed than cultural vocabulary. The analysis also extends to a comparison of loanwords from vocabulary in general versus basic lexicon, which reveals basic vocabulary to be at least *two and a half times less likely* to be borrowed than words of the general lexicon (Brown 2017:273–274).

Metrics are available for determining the extent to which the results of quantitative screening support the genealogical relationship of compared languages, two of which incorporate basic vocabulary. These metrics are the *number* of screened-in sets of a language comparison, each of which entails basic vocabulary (*NBV*) and the *percentage* of all screened-in sets for a language comparison that entail basic vocabulary (*BV%*). A comparative set involves basic vocabulary if any word of the set is glossed by a word designating a Swadesh basic-vocabulary item or if some reflex of a proto-word of a set is so glossed (Brown 2017:274). Both metrics, but especially *BV%*, gauge the extent to which observed lexical similarity is due to kinship or borrowing. Putting aside *NBV* for the moment, a smaller *BV%* value is more strongly associated with borrowing, and a larger *BV%* more strongly with kinship.

(For comparisons involving 100 or fewer quantitatively screened-in sets, *BV%* is the number of sets involving basic vocabulary divided by the total number of sets observed for a comparison, multiplied by 100. For a collection of comparisons with more than 100 sets, *BV%* is the absolute number of sets involving basic vocabulary [i.e., its value is the same as that of *NBV*].)

1.2.3 ASJP

While developmentally distinct from the quantitative screening method described above, ASJP is another metric for quantificational detection of language relationships. ASJP is designed for automated binary comparison of word lists (almost 10,000 in number available) from most of the world's languages and dialects (Brown et al. 2008, Holman et al. 2008, The ASJP Database (clld.org)). These word lists consist of basic vocabulary on Swadesh's (1955a) 100-item list, from which a shortened version comprising the forty most stable words is predominantly used. ASJP values provide a measurement of the degree to which two languages are lexically similar. The more lexically similar two compared languages are found through ASJP, the more closely related these languages are assumed to be. Since items compared involve only basic vocabulary, lexical similarities are assumed to be less likely the products of borrowing than otherwise. Values of ASJP range from near or just below 0 to 100. Compared lists with an index of 100 are identical. Compared lists with values below 100 are less and less alike as their values approach zero.

In using ASJP, word lists are transcribed into a standardized simplified orthography called ASJPcode. (In contrast, quantitative screening uses the original orthography of sources.) A computer determines the phonological similarity between pairs of lists using an algorithm based on Levenshtein distance (also known as edit distance) (Levenshtein 1966). When fed into algorithms such as neighbor-joining (Saitou and Nei 1987), the derived pairwise similarities produce branching trees for the languages resembling family trees of the traditional comparative method (e.g., figures 5.2 and 5.4). Such trees consistently show similarity to classifications provided by expert human comparativists. A growing collection of ASJP word lists has been posted online in successive versions, of which the latest ("version 20," Wichmann et al. 2022) contains 10,168 lists from 5,590 languages with different ISO639-3 codes. (Dates for which pairs of possibly related languages have separated from one another can be estimated based on Levenshtein distances for ASJP languages [Holman et al. 2011]. ASJP Chronology is discussed in chapter 9 [9.2.].)

1.2.4 QUANTITATIVE SCREENING AND ASJP COMBINED

The combination of these two approaches provides a more complete estimate of evidence strength for a language relationship than either approach on its own. We combine the two into a single approach, in which *NBV* of quantitative screening and *ASJP* are metrics with values merged to render a single evaluation for language relationships. If a connection is detected by this merger, we can then consult the *BV%* of quantitative screening to determine whether the connection involves language kinship or language contact (borrowing).

This combined methodology is illustrated in table 1.1, providing comparisons and evaluations for thirty-two pairs of widely distributed languages. These are extracted from 232 comparisons given in online appendix A. Comparison selections from online appendix A included in table 1.1 involve the first four pairs listed in each of the eight sections identified by strength of evidence; for example, pairs 1 Defaka/Nkọrọ, 2 Marithiel/Maranunggu, 3 Chambri/Kopar, and 4 Alamblak/Kapriman, evidence for each of which is identified as *Strong for Kinship*.

Column 1 (left-most) of table 1.1 gives ID numbers for each of the thirty-two language comparisons of column 2. Possible language-family membership for each comparison is found in column 3. Column 4 supplies the broad-area location of languages/families.

Metrics *BV%, NBV,* and *ASJP,* respectively of columns 7, 8, and 9, are discussed above in sections 1.2.2 and 1.2.3. Values supplied for *NBV* and *ASJP* in their respective columns are either *underlined* or not *underlined*. *Underlined* values for *NBV* and *ASJP* are all greater than or equal to average, and non-*underlined* values for these are less than average. Averages for these

1.1 Thirty-Two Language Comparisons Evaluated Through Combined Quantitative Screening and ASJP Approaches

ID	*Language Comparison*	*Possible Group*	*Area*	*Sets*	*Cor*	*BV%*	*NBV*	*ASJP*	*Evidence for Degree of Support*
1	Defaka/Nkọrọ	Ijoid	A	25	15	*100*	*25*	*34.95*	Strong for Kinship
2	Marithiel/Maranunggu	Western Daly	P	26	17	*100*	*26*	*30.59*	Strong for Kinship
3	Chambri/Kopar	Lower Sepik	P	18	8	*55.6*	*10*	*24.75*	Strong for Kinship
4	Alamblak/Kapriman	Sepik Hill < Sepik	P	37	19	*56.8*	*21*	*21.38*	Strong for Kinship
37	Proto-Awyu-Dumut/ Proto-Mountain OK	Greater-Awyu-Greater Ok < TNG	P	28	14	*50*	*14*	2.6	Moderate for Kinship
38	Takelma/Proto-Kalapuya	Plateau Penutian	N	17	10	*70.5*	*12*	−0.21	Moderate for Kinship
39	Proto-Paman/Proto-Ngayarda	Pama-Nyungan	P	28	13	*71.4*	*20*	3.14	Moderate for Kinship
40	Proto-Popolocan/Isthmus Zapotec	Otomanguean	M	31	15	*32.3*	*10*	2.82	Moderate for Kinship
79	Takelma/Yokuts	Penutian	N	3	2	*66.7*	2	2.43	Weak for Kinship
80	Kire/Tangu	Lower Ramu	P	6	3	*66.7*	4	2.01	Weak for Kinship
81	Awtuw/Abau	Sepik	P	3	2	*66.7*	2	1.93	Weak for Kinship
82	Jicaque/Comecrudo	Hokan	N & M	11	8	*64.6*	6	3.32	Weak for Kinship
106	Atakapa/Chitimacha	Gulf	N	67	26	19.4	*13*	*5.4*	Strong for Borrowing
107	Shasta/Karok	Hokan	N	71	33	26.8	*19*	*3.83*	Strong for Borrowing
108	Resígaro/Bora	Resígaro-Boran	S	30	14	26.7	*8*	*13.16*	Strong for Borrowing

109	Proto-Shuar/ Candoshi	Jivaroan-Candoshi	S	60	22	15	*9*	*10.73*	Strong for Borrowing
112	Burushaski/Indo-European	Burushaski-Indo-European	E	75	30	24	*18*	*2.17*	Moderate for Borrowing
113	Yana/Karok	Hokan	N	27	15	25.9	*7*	2.1	Moderate for Borrowing
114	Proto-Mixtecan/ Amuzgo	Otomanguean	M	57	26	28.1	*16*	0.98	Moderate for Borrowing
115	Proto-Eskimo/Aleut	Uralo-Siberian	N	107	28	16	*16*	0.9	Moderate for Borrowing
126	Kujarge/Birgit	East Chadic	A	0	0	0	0	*9.98*	Moderate for Connection
127	Yuki/Yawelmani Yokuts	Yukian-California Penutian	N	0	0	0	0	*3.68*	Moderate for Connection
128	Proto-Uralic/Proto-Yukaghir	Uralo-Siberian	E	0	0	0	0	*3.65*	Moderate for Connection
129	Kwoma/Yessan	Sepik	P	0	0	0	0	*16.79*	Moderate for Connection
145	Wintu (Wintun)/ Southern Sierra Miwok	California Penutian	N	32	21	12.5	4	3.13	Weak for Borrowing
146	Proto-Popolocan/ Proto-Mixtecan	Otomanguean	M	18	8	16.7	3	3.28	Weak for Borrowing
147	Zuni/Yawelmani Yokuts	Zuni-California Penutian	N	13	10	23.1	3	2.75	Weak for Borrowing
148	Natchez/Proto-Muskogean	Gulf	N	44	23	13.6	6	1.64	Weak for Borrowing

(*Continued*)

Table 1.1 (Continued)

ID	*Language Comparison*	*Possible Group*	*Area*	*Sets*	*Cor*	*BV%*	*NBV*	*ASJP*	*Evidence for Degree of Support*
162	Proto-Karajá/Ofaye	Macro-Je	S	0	0	0	0	3.21	No Support for Connection
163	Paya/Talamancan	Chibchan	M	0	0	0	0	3.2	No Support for Connection
164	Proto-Je (Proto-Cerratense)/ Proto-Karaja	Macro-Je	S	0	0	0	0	3.13	No Support for Connection
165	Muisca/Paya	Chibchan	M	0	0	0	0	3.02	No Support for Connection

Source: Generated by the authors.
Notes: A = Africa, E = Eurasia, M = Middle America (Mexico plus Central America), N = North America, S = South America, P = Papua New Guinea and Australia, TNG = Trans-New Guinea.
ASJP = ASJP index of lexical similarity for two compared languages.
BV% = Percentage of basic-vocabulary items of all screened-in comparative sets for a language comparison.
NBV = Number of screened-in comparative sets that involve basic vocabulary.
Italicized values of *BV%*, NBV, and ASJP are either equal to or greater than mean value (mean value for *BV%* = 28.42, for NBV = 6.6, for ASJP = 3.36). If the *BV%* value is 0, it is not *underlined*. (Percentages are based on results found for the 232 comparative sets of online appendix A.)

are calculated from information provided for the 232 comparisons in online appendix A. The means for *BV%*, *NBV,* and *ASJP* respectively are 28.42, 6.61, and 3.36.

The final column of the table strays away from calculations based on numbers to speculation based on numbers. For example, if values for *NBV* and *ASJP* for a language comparison are both greater than their respective means (see pairs 1, 2, 3, 4, 106, 107, 108, 109), evidence for the comparison is *Strong* or, in other words, robustly supports the connection of the two compared languages. In addition, if its *BV%* value is also greater than average (*underlined*), evidence for the comparison is *Strong for Kinship.* In contrast, *NBV* and *ASJP* values are both less than their means (not *underlined*) for comparisons 79, 80, 81, 82, 145, 146, 147, and 148, indicating *No Support for Connection* or, in other words, no support for any historical association, be it through kinship or borrowing. The remaining six interpretations of table 1.1 follow (with annotation):

Moderate for Kinship: One *underlined* value and one non-*underlined* for *NBV* and *ASJP*, where the value for *BV%* is greater than its mean and *underlined.* Compared languages with this evaluation may or may not be genealogically affiliated, but the evidence is strong enough to warrant further study.

Weak for Kinship: Neither *NBV* nor *ASJP* is *underlined*, but the *BV% value* is greater than its mean. The relationship is not likely. Evidence is not strong enough for recommending further investigation.

Strong for Borrowing: Both *NBV* and *ASJP* are *underlined*, with the value for *BV%* being smaller than its mean. The two compared languages are likely to be related through borrowing either directly or indirectly. Indirect borrowing involves a third-party, intermediate language in contact with both of the other two.

Moderate for Borrowing: *NBV* is *underlined*, or *ASJP* is *underlined*, but not both, and the *BV%* value is less than its mean. Compared languages may or may not have a relationship of borrowing, either direct or indirect. Evidence is strong enough to encourage further study.

Weak for Borrowing: Neither *NBV* nor *ASJP* is *underlined*, and *BV%* is less than its mean but still greater than zero (value > 0). An affiliation is unlikely. Evidence is not strong enough to recommend additional study.

Moderate for Connection: Only *ASJP* is *underlined* or only *NBV* is *underlined*, and the *BV%* value is zero (0). Compared languages may or may not be either genealogically affiliated or related through borrowing. Evidence is strong enough to recommend further attention.

The key to distinguishing a group of borrowed words from a group of non-borrowed words in the system is *BV%*. As noted in 1.2.2, a smaller *BV%* is associated more strongly with borrowing, and a larger *BV%* more strongly with kinship. If the value of a comparison's *BV%* is found to be less than average (<28.42), this is indicative of lexical similarities in the comparison most likely due to borrowing. Conversely, when the *BV%* value is equal to or greater than average (≥28.42), this indicates resemblance due to a genealogical relationship. These conditions only hold when the value of *BV%* is not zero (≠ 0).

Chapter 2

Evaluation of Aleut-Utian Comparative Evidence

2.1 INTRODUCTION

In this chapter, we apply both quantitative and traditional methods to evaluate comparative support for the proposal that Aleut (Alaska) and Utian (California) are genealogically related.

Online appendix A for chapter 1 presents results for a quantitative evaluation of comparative evidence for 232 lexical sets, among which is an assessment for Aleut/Proto-Miwok (comparison 56). If Proto-Miwok is genealogically related to Aleut, as we propose, then Proto-Utian, the immediate parent language of Proto-Miwok, is also related to Aleut. This relationship would, in turn, support our main proposal, which asserts that Utian languages of central California and the Aleut language of Alaska are genealogically affiliated.

Our study began with an investigation of the possible relationship between Aleut and Proto-Utian using the quantitative approach described in 1.2. While the quantitative evidence for this relationship provides *No Support for Connection*, this result is not particularly surprising since a genealogical connection between Aleut and Proto-Utian would necessarily be chronologically remote. As a result, much supportive comparative evidence may well have been eroded by the waves of time (attrition).

Another factor possibly helping to explain the dearth of positive Aleut/Proto-Utian comparisons involves vocabulary for Costanoan used in Callaghan's (2014) Utian dictionary. Each Proto-Utian lexical reconstruction in Callaghan requires both Miwok and Costanoan cognates as supporting evidence (see 9.4.4). Unfortunately, unlike the circumstance for Miwok languages, no full-fledged dictionaries are available for Costanoan languages.

Callaghan herself served as lexicographer for four Miwok dictionaries: Lake Miwok (1965), Bodega Miwok (1970), Plains Miwok (1984), and Northern Sierra Miwok (1987). Costanoan sources are much shorter vocabulary lists, prepared mainly by priests at California missions during the nineteenth century and reproduced by scholars in the twentieth century, for example, de la Cuesta (Mason 1916) and Alphonse Pinart (Heizer 1952).

Nevertheless, we investigated further by applying our quantitative method to the evaluation of evidence for an Aleut/Proto-Miwok relationship, assuming that the more extensive data for Aleut-Proto-Miwok would show less attrition of positive comparative evidence, if indeed a Proto-Aleut-Miwok relationship were valid. The strength of evidence for this comparison (no. 56 of the 232) is assessed as *Moderate for Kinship,* indicating, "Compared languages with this evaluation may or may not be genealogically affiliated, but the evidence is strong enough to warrant further study" (1.2.4).

Section 2.3 of this chapter describes details of the application of our quantitative approach to Aleut/Proto-Miwok. This analysis is followed, in section 2.4, by further study through application of the traditional method of comparative linguistics to the Aleut-Utian proposal. The latter involves identifying cognates (two or more words derived from the same ancestral word), sound correspondences, and reconstructing a phonological inventory (2.4.2) and a short vocabulary (2.4.6) for Proto-Aleut-Utian (PAU).

2.2 ALEUT AND UTIAN

Aleut is a single language of southwest Alaska spoken by Native Peoples of the Aleutian, Pribilof, and Commander Islands (Alaska) (see map, figure 2.1). Utian is a small language–family of north-central California with languages historically spoken by groups geographically distributed around San Francisco Bay and extending into the interior (see map, figure 2.2).

2.2.1 Aleut

Aleut (also called Unangam) has three dialects. Aleut words used in comparative sets assembled here are from Bergsland's (1994) *Aleut Dictionary.* Languages of the Eskimo family (figure 2.1) are long regarded as the closest relatives of Aleut; together they comprise the Eskimo-Aleut family. In this study, the quantitative approach does not support the genealogical affinity of Aleut and Eskimo, but is instead is supportive of extensive borrowing between the two. Our quantitative comparison provides evidence that the closest kin of Aleut are Utian languages of California.

Figure 2.1 Distribution of Aleut and Eskimo Languages in West Alaska. *Source*: Adapted from *Federal Field Committee for Development Planning in Alaska.*

2.2.2 Utian

Utian is a language–family of north-central California with two major branches, Costanoan (*aka* Ohlone) and Miwok. No speakers of Costanoan languages now survive, and only a few individuals speak a Miwok language. While tribal elders who were fluent speakers of Costanoan languages passed away in the twentieth century, some Ohlone people are relearning their languages today, and some Miwok tribes have retained elements of their languages in historic times and are similarly undertaking relearning.

Callaghan (2014:8, 14–15) in her book, *Proto Utian Grammar and Dictionary with Notes on Yokuts*, the main source of Utian words used in our comparative sets (2.4.6), gives the following classification for the Utian family, also supplying in parentheses language-name abbreviations used in her dictionary and adopted by us in this work (see Abbreviations and Conventions, 2.4.1).

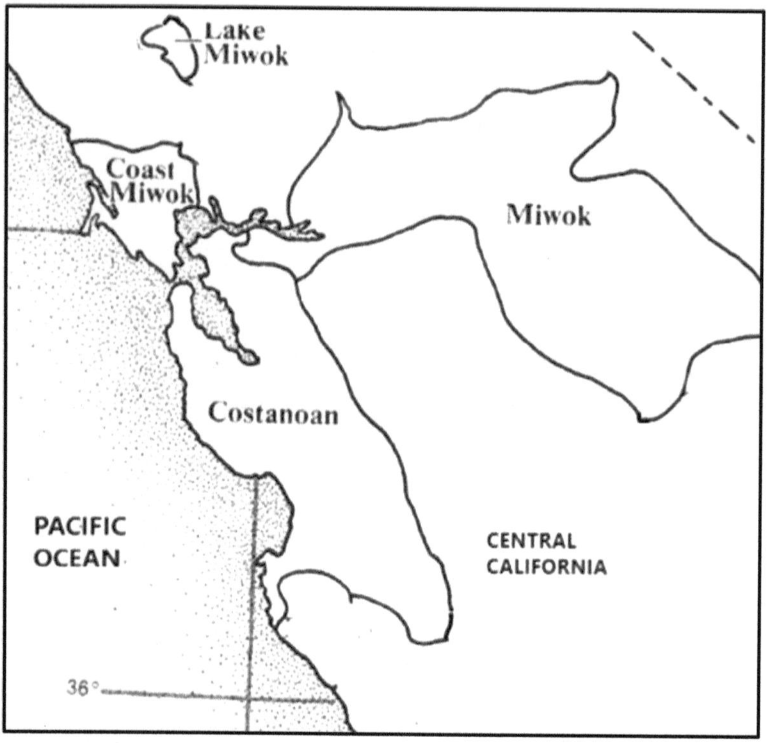

Figure 2.2 Distribution of Historic Utian languages (Costanoan and Miwok) in Coastal Central California. *Source*: Adapted from Dixon and Kroeber (1919).

Miwok (Mi)
Eastern Miwok (Mie)
Sierra Miwok (Mis)
Southern Sierra Miwok (Miss)
Central Sierra Miwok (Mics)
Northern Sierra Miwok (Mins)
Plains Miwok (Mip)
Saclan (Misac)
Western Miwok (Miw)
Coast Miwok (Mic)
Bodega Miwok (Mib)
Costanoan (Co)
Northern Costanoan (Con)
Southern Costanoan (Cos)
Mutsun-San Juan Bautista (Csjb)
Rumsen (Cru)

Callaghan (2014) reconstructs words for Proto-Utian (PU), Proto-Miwok (PMi), and Proto-Costanoan (PCo), as well as for proto-languages subordinate to these, such as Proto-Southern Sierra Miwok (PMiss) and Proto-Southern Costanoan (PCos). All historical-language reflexes (descendant words) of reconstructed proto-words are presented in her work. Words reconstructed for PU, PMi, and PCo are the primary Utian lexical materials used in our comparative sets, but sets also include secondary references to words from other Utian offspring proto-languages and from individual Utian languages (most extracted from Callaghan 2014; for other sources see 2.4).

Three groups neighboring Utian in California—Wintu, Maidu, and Yokuts—have been proposed as the closest kin of Costanoan and Miwok (Utian) languages. Together, these five groups have been (and still are, in some quarters) considered constituents of the California Penutian family (Dixon and Kroeber 1919). Our application of the quantitative approach to California Penutian finds some support for this group (chapter 5, 5.3). However, Callaghan (2014), who rejects California Penutian, instead considers Utian to be genealogically related only to Yokuts and assembles comparative sets for this proposal in her book. Applying the quantitative approach to these sets shows little support for the Utian/Yokuts proposal (chapter 5). In contrast, both quantitative and traditional comparisons, especially the latter, find support for Aleut/Proto-Miwok.

2.3 QUANTITATIVE METHOD APPLIED TO ALEUT/PROTO-MIWOK

The first step in quantitative evaluation is to obtain an assemblage of two-language lexical comparisons, typically using published comparative sets compiled by linguists (1.2). If a compilation of binary comparisons for two languages is not available, an alternative strategy is the self-production of sets using sources such as available dictionaries, vocabularies, and word lists. No collection of binary comparisons for Aleut/Proto-Miwok was available when we began our evaluation of these languages, so we compiled our own from dictionaries, most notably from Bergsland's (1994) dictionary for Aleut and Callaghan's (2014) dictionary for Proto-Utian, both of which are large lexicons for the respective languages.

These dictionaries were searched for lexical "look-alikes" for Aleut and Miwok. Look-alikes are words from two or more languages showing resemblance in both sound and meaning and are typically used in comparative sets for evaluation by the traditional method. We initially compiled somewhere between 275–300 binary look-alikes for Aleut and Miwok.

Our quantitative analysis proceeds with the evaluation of lexical comparisons using a combination of quantitative screening and ASJP approaches (1.2.4). After iterations in which all unlikely Aleut/Miwok comparative sets and sound correspondences were screened out one-by-one (1.2.1.4), twenty-eight screened-in comparative sets and 13 correspondences remained (comparative sets are given below in this section, and the thirteen screened-in sound correspondences are reported in table 2.1).

28 Screened-in comparative sets for Aleut/Proto-Miwok
1. A *tani-χ* "forehead"
PMi **ṭik:i(t)* "forehead"
Evaluation points: 5 (4,1,0)

2. A *takaχ* "woman's braid," *takaχ-six* "to make a braid"
PMi **ṭaṭ-* "braid hair"
Evaluation points: 5 (4,1,0)

3. A *tayaʁu-χ* "man, human being; man, male human"
PMi^{a} **ṭayi-* "man" (PU **ṭa(:)li-ṣ* "man [male]"; R: Mim *tayiṣ* "man [male]," Mib *tayih* "man [male]")
Evaluation points: 9 (8,1,0) BV

4. A *tut-* "to hear, get to hear"
PMi^{a} **ṭole:-* "to hear" (PU **ṭol-* "to hear"; R: PMis *ṭole:-yɨ* "to hear, understand")
Evaluation points 7 (6,1,0) BV

5. A *tuɣi-l* "to be dotted," A *tuɣi-χ* "spot, dot"
PMi^{a} *ṭona-* "spotted" (PU **ṭona-* "spotted"; R: PMie **ṭona-ṭ-:a-* "spotted, speckled")
Evaluation points: 5 (4,1,0)

6. A *čat-* "to get partially dried (of fish to be dried)"
PMi **čuṭul* "dry"
Evaluation points: 5 (4,1,0) BV

7. A *kita-χ* "foot (of person, bird, reindeer)"
PMi^{a} **kolo* "foot" (PU **kolo* :"foot"; R: PMie **kolo-* "foot")
Evaluation points: 8 (6,1,1) BV

8. A *kitu-χ* "louse, nit"
PMi **ket:ɨ-* "head louse"
Evaluation points: 5 (4,1,0) BV

9. A *-kuča-* "small, small-sized" (form is a suffix)
PMi^{a} **kuč:i* "small" (PU **kuč-i* "small, nice"; R: Mics kuci- "good")
Evaluation points: 6 (4,1,1) BV

10. A *-qa-* "remote tense: remote past or, with certain auxiliaries . . . , remote future"
PMi **ka:* ~ **ka* "past tense marker" (R: Mip *-ka-* "vs remote past")
Evaluation points: 6 (4,1,1)

11. A *qaka-lix* "to be dry," *qakat-* "to dry, to make dry (e.g., clothes, guts)"
PMi **ka:*y "dried up"
Evaluation points: 5 (4,1,0) BV

12. A *qaŋa-lix* "to be done, cooked . . . to ripen (of berries . . . of corn, seeds . . .)"
PMi **kili* "ripe, well done" (R: Mics *kil-ŋe-* "to be cooked")
Evaluation points: 6 (4,1,1)

13. A *qahmix, qamiɣasi-χ* "door"
PMi **ka:* "door, i.e., a basket covering the opening to a house" (R: PMiw **ka:* "door, basket")
Evaluation points: 5 (4,1,0)

14. A *hataʁičikna-χ* "pale," *'haʁaði-l* "to be pale"
PMi **poto(k)* "gray, gray-haired" (R: Mil *poto:-poto* "gray, pale")
Evaluation points: 5 (4,1,0)

15. A *hani-χ* "lake" (Note: *aniiða-χ* "pond," *haanimiklaaðaaluχ* "pool, bath house")
PMi **pol-ku* "to flood something" (R: PMis **pol:ay-* "pond, puddle, pool")
Evaluation points: 5 (4,1,0)

16. A *huluχ-six* "to blink, wink (of eyes)"
PMi **pel-* "pertaining to eyes" (PU **pel-* "pertaining to eyes"; R: Mil *pel-ak* "to blink")
Evaluation points: 5 (4,1,0)

17. A *ilɣa-lix* "to search, to look for"
PMi[a] **wel-* "to look for something" (PU **wele* "to look for"; R: PMis **wel-ṣɨ* "to look for something")
Evaluation points: 5 (4,1,0)

18. A *ilɣut-* "to receive, get (object, command…) . . . to take (wife)"
PMi **wel-ki* "to get, to take"
Evaluation points: 5 (4,1,0)

19. A *ikalqu-χ* "ring finger, fourth finger"
PMi **ʔek:u-s* "hand, finger" (PU **ʔek:u* "hand")
Evaluation points: 7 (6,1,0) BV

20. A *iqut-* "to break"
PMi **koṭ:*a "to break something, to break something in two" (R: Mics *koṭ-ku-* "to break a limb")
Evaluation points: 6 (6,0,0)

21. A *at-* "lower part, area below, underside, bottom"
PMi **ʔa:la* "below, east" (R: Miss *ʔal:a-* "underside," Mics *ʔal:a-* "down, low")
Evaluation points: 5 (4,1,0)

22. A *aliiʁi-χ* "wolf" (Canis)
PMi **ʔole* "coyote" (Canis)
Evaluation points: 6 (6,0,0)

23. A *aya-* "friend, cousin"
PMi **ʔoy:am* "friend, partner, in-law"
Evaluation points: 5 (4,1,0)

24. A *ayaʁi-lix* "to be ashamed, to be bashful"
pMi[a] **ʔuya-* "bashful, ashamed" (PU **ʔuy(:)a* "ashamed"; R: PMiw **ʔoya-ṣ* "bashful, ashamed")
Evaluation points: 7 (6,1,0)

25. A *ayxa-lix* "to exceed, be more or greater etc. than"
PMi **ʔa:y* "again, another, more," *ʔa:y-ɨ-* "to increase"
Evaluation points: 5 (4,1,0)

26. A *ulax* "enclitic negative"
PMi **hel:a* "no, not"
Evaluation points: 5 (4,1,0) BV

27. A *uluuða-χ* "red"
PMi[a] *yoli-* "red" (PU **yoli* "red"; R: Mics *yol-ol:i* "red, a red one")
Evaluation: 5 (4,1,0) BV]

28. A *uya-lix* "to call, to send for, to invite"
PMi **yay:u* "to call, invite"
Evaluation points: 6 (4,1,1)

In the preceding collection of twenty-eight comparative sets on which the thirteen sound correspondences of table 2.1 are based, an Aleut word is either compared with a Proto-Miwok word reconstructed by Callaghan (PMi, 2014) or with one reconstructed by the current authors (PMi[a]). We reconstruct PMi[a] when Callaghan fails to provide a PMi word but nevertheless reconstructs both a PU word and a Miwok reflex of the PU word (other than its PMi form). The reconstruction of the PMi[a] reflex by the authors is based on the comparative phonology of observed forms as given by Callaghan (2014), and on the critical assumption that if a PU word shows a Miwok-language reflex (other than PMi), a PMi form necessarily pertained to the history of the lexical set.

Table 2.1 Thirteen Quantitative Screened-In Aleut/Proto-Miwok Sound Correspondences

Aleut	*Proto-Miwok*	*Supporting Sets*
h	*p*	|14,|15,|16,|
t	*ṭ*	|1,|2,|3,|4,|5,|6,|20,|
t	*l*	|4,|7,|21,|
l	*l*	|16,|17,|18,|22,|26,|27,|
k	*k*	|7,|8,|9,|19,|
q	*k*	|10,|11,|12,|13,|20,|
y	*y*	|3,|23,|24,|25,|28,|
i	*e*	|8,|17,|18,|19,|22,|
a	*a*	|2,|3,|10,|11,|13,|21,|23,|24,|25,|26,|
a	*i*	|1,|3,|9,|12,|12,|
a	*u*	|6,|19,|24,|28,|
a	*o*	|7,|14,|15,|22,|23,|
u	*o*	|4,|5,|20,|27,|

Source: Generated by the authors.

Table 2.2 Results of Combined Quantitative Screening and ASJP Applied to Aleut/Proto-Miwok Comparative Sets

Result Type	*Value*
Quantitatively screened-in sets (*Sets*)	28
Screened-in correspondences (*Cor*)	13
Basic-vocabulary percentage (*BV%*)	*35.7 (10/28 × 100)*
Number of screened-in sets with basic vocabulary (*NBV*)	*10*
ASJP similarity	−1.14
Strength of support	*Moderate for Kinship*

Source: Generated by the authors.

Table 2.2 gives results of the quantitative approach applied to Aleut/Proto-Miwok. Values of *BV%* and *NBV* are *underlined*, indicating these are above average: mean *BV%* = 28.42, mean *NBV* = 6.6 (based on *BV%* and *NBV* values of the 232 comparisons in online appendix A). Since the value of *BV%* is above average, observed lexical similarity for Aleut and Proto-Miwok is likely due to kinship rather than to borrowing (1.2.2). Among *BV%*, *NBV*, and *ASJP*, only *ASJP* is below average (mean *ASJP* = 3.36, based on the *ASJP* values of the 232 comparisons in online appendix A). If *ASJP* were instead greater than average, all three values would exceed their means and evaluation for Aleut/Proto-Miwok would be *Strong for Kinship*. Since only one of the two metrics (*NBV* and *ASJP*) shows an above-average value, quantitative evaluation of Aleut/Proto-Miwok with *Moderate for Kinship* (1.2.4) is one rank down from full strength (*Strong for Kinship*).

Other possible evaluations of Aleut/Proto-Miwok might have been either *Strong for Kinship* or *Weak for Kinship*. If the former, a genealogical relationship is considered highly likely with no further research being required for validation. If the latter, a genealogical affiliation is considered highly improbable with no further research likely to change findings. Since the result is *Moderate for Kinship*, additional research applied to Aleut/Proto-Miwok lexical evidence may show support for a genealogical relationship and is therefore worthy of pursuit using the traditional comparative method, as follows.

2.4 TRADITIONAL APPROACH APPLIED TO ALEUT/UTIAN

Sixty-six comparative sets and 26 sound correspondences for Aleut and Utian languages are found through the use of the traditional approach. As in the quantitative approach, principal lexical sources for sets are Bergsland (1994) for Aleut and Callaghan (2014) for Miwok. Two other Utian sources are also used: Collier and Thalman (1991) and Thalman (1993). These two sources are important for their abundance of Coast Miwok words, especially for coastal and marine referents (e.g., seaweed, kelp, clams, sea anemone, whales) not found in Callaghan (2014). A few other sources are also used, especially for Costanoan lexical items: Heizer (1952, 1955), Kroeber (1910), and Mason (1916).

Reconstructed for each comparative set is a PAU word from which member words of the set (reflexes) are presumed to have developed. (For a straightforward, step-by-step explanation of how proto-words [in general] are reconstructed, we recommend consultation of Campbell [1999:108–162]). The order of presentation of comparative sets is determined by the first sound segment of reconstructed PAU words. The order accords with the way in which segments are generally organized in linguistic studies, i.e., by positions in the oral apparatus where they are produced, with frontal segments first, dorsal segments final, and others in between, followed by vowels. The order used here is:

p, m, w, t, ṭ, č, s, ṣ, n, l, r, k, q, ʁ, y, ʔ, h, i, ɪ, e, æ, ɨ, a, u, ʊ, o, ɔ.

2.4.1 Abbreviations and Conventions

A = Aleut, Later Aleut
AAY = Alutiiq Alaskan Yupik (Eskimo)
AU = Aleut-Utian
BV = basic vocabulary (Swadesh [1955a] 100-item list)
C = Consonant

CAY = Central Alaskan Yupik (Eskimo)
Cru = Rumsen (Monterey, Carmel Costanoan)
Csjb = Mutsun (San Juan Bautista Costanoan)
Den = Dena'ina
EA = Early Aleut
Mib = Bodega Miwok
Mics = Central Sierra Miwok
Mil = Lake Miwok
Mim = Marin Miwok
Mins = Northern Sierra Miwok
Mip = Plains Miwok
Miss = Southern Sierra Miwok
PAU = Proto-Aleut-Utian
PE = Proto-Eskimo
PCo = Proto-Costanoan reconstructed by Callaghan (2014)
PCo^a = Authors' reconstruction of Proto-Costanoan
PMi = Proto-Miwok reconstructed by Callaghan (2014)
PMi^a = Authors' reconstruction of Proto-Miwok
PMie = Proto-Eastern Miwok
PMis = Proto-Sierra Miwok
PMiw = Proto-Western Miwok
PU = Proto-Utian reconstructed by Callaghan (2014)
PU^a = Authors' reconstruction of Proto-Utian
PWintu = Proto-Wintu
PX^a = Authors' reconstruction (where X is a place holder for any language abbreviation)
R = Reflex
V = Vowel
#X = Stem-initial sound segment (where X is any sound segment or Ø)
–X = Non-stem-initial segment (where X is any sound segment)
[. . .] Brackets, other than those applied within parentheses, encompass original orthography
Ø = Nil, no-sound segment

2.4.2 Phonological Inventories and Correspondences

Phonological inventories for both Aleut (table 2.4) and Proto-Utian (table 2.5) are mostly derived from the inventory of Proto-Aleut-Utian (table 2.3). PAU consists of seventeen consonants and ten vowels. The Aleut inventory consists of twenty consonants and three vowels, sharing twelve consonants and three vowels with PAU. Proto-Utian, with fifteen consonants and six vowels, shares thirteen consonants and six vowels with PAU. A total of

Table 2.3 Phonological Inventory for Proto-Aleut-Utian Based on the Sixty-Six Comparisons Presented in 2.4.6

	Proto-Aleut-Utian Consonants					
Type	*Labial*	*Dental*	*Palatal*	*Velar*	*Uvular*	*Glottal*
Stop	*p*	*t* | *ṭ*		*k*	*q*	*ʔ*
Affricate			*č*			
Fricative			*s* | *ṣ*		*ʁ*	*h*
Nasal	*m*	*n*				
Approximant	*w*	*l*	*y* | *r*			

Notes: Segments with sub-dots (*ṣ*) are retroflexed. Consonant chart is organized by "manner of articulation" (left column) and by "place of articulation" (top row). Vowel chart is organized by the "height of tongue" (column) and the "position of the tongue (front, mid, back)" (row).

Proto-Aleut-Utian Vowels		
i	*ɨ*	*u*
ɪ		*ʊ*
e		o
æ	a	ɔ

Source: Generated by the authors, with vowels based on Pullum and Ladusaw (1986, 260).

Table 2.4 Aleut Phonological Inventory

	Aleut Consonants					
Type	*Labial*	*Dental*	*Palatal*	*Velar*	*Uvular*	*Glottal*
Stop	*p > h*	*t*		*k*	*q*	
Affricate			*č*			
Fricative		ð	*s* *z**	*x* *ɣ*	*χ* *ʁ*	*h*
Nasal	*m*	*n*		*ŋ*		
Approximant	*w*	*ɬ* *l*	*y*			

*Palatal-alveolar sibilant found only in the Atkan dialect.

Aleut Vowels (Can Be Long or Short)		
i		*u*
	a	

Source: Modified from Bergsland (1994, xvii).

Table 2.5 Proto-Utian Phonological Inventory

	Proto-Utian Consonants				
Type	*Labial*	*Dental*	*Palatal*	*Velar*	*Glottal*
Stop	*p*	*t* | *ṭ*		*k* | *kʷ*	*ʔ*
Affricate			*č*		
Fricative			*s* | *ṣ*		*h*
Nasal	*m*	*n*			
Approximant	*w*	*l*	*y*		

Note: Sub-dot indicates retroflexion.

Proto-Utian Vowels		
i	*ɨ*	*u*
e		*o*
	a	

Source: Modified from Callaghan (2014, 69–70).
Note: Both Utian consonants and vowels can be long or short depending on context. Consonant and vowel length does not reconstruct for PAU.

eleven consonants and three vowels are shared by all three systems: *p, m, w, t, n, l, č, s, y, k, h* and *i, a, u*. Historic Aleut lacks *p*, but this is the result of a recent sound change in which all Early Aleut *p*'s shifted to Later Aleut *h*.

Of the three languages of tables 2.3, 2.4, and 2.5, inventories for PAU (table 2.3) and Aleut (table 2.4) are most different from one another. Aleut shows seven consonants not found in PAU (nor found in PU for that matter). These are *ð, ł, z, x, ɣ, ŋ,* and *χ,* (*z* occurs in only one of the Aleut dialects). PAU shows four consonants not found in Aleut: *ṭ, ṣ, r*, and *ʔ*. While the latter four segments could have simply been lost by Aleut in its development from PAU, the seven consonants found in Aleut but not in PAU must have been added. This addition probably involved contact with and borrowing from some non-PAU language or languages. The most likely source language was Proto-Eskimo and, later, one or more of the latter's daughter languages. Table 2.6 presents the phonological inventory for Proto-Eskimo reconstructed by Fortescue et al. (2010).

Of the seven Aleut consonants not found in PAU, the reconstructed Proto-Eskimo inventory shows four: ð, *ł, ɣ,* and *ŋ*. As mentioned above, *z* is found in only one of the Aleut dialects and, thus, is likely a late addition. Fortescue et al. (2010) include both *x* and *χ* in their Proto-Eskimo phonological chart, but in parentheses. Parentheses indicate their non-Proto-Eskimo phoneme status, but also their presence in early Proto-Eskimo daughter languages such as Proto-Yupik Eskimo (Fortescue, pers. com., 2022). Aleut could have borrowed *x* and *χ* from these later Proto-Eskimo offspring.

Table 2.6 Phonological Inventory for Proto-Eskimo

	Proto-Eskimo Consonants				
Type	*Labial*	*Dental*	*Palatal*	*Velar*	*Uvular*
Stop	*p*	*t*		*k*	*q*
Affricate			*č*		
Fricative	*v*	*ð*		*ɣ*	*ʁ*
Nasal	*m*	*n*		*ŋ*	
Approximant		*ɬ* *l*	*y*		

Proto-Eskimo Vowels		
i	*ɨ*	*u*
	ə	
	a	

Source: Generated by the authors, based on Fortescue et al. (2010).

A competing explanation is that Aleut and Proto-Eskimo may share *ð, ɬ, ɣ,* and *ŋ* due to their mutual inheritance of these phonemes from a possible common parent, Proto-Eskimo-Aleut (but whose existence we challenge with robust counter-evidence; see especially chapter 5, 5.5).

2.4.3 Context-Conditioned Change

Table 2.7 presents correspondence series (across PAU, A, PU, PMi, and PCo) among which some involve conditioned change. For example, in the correspondence series for PAU **w*, the Aleut reflex is realized as *y* in all phonological contexts except stem initially (#__) where it is nil (Ø), that is, representing no-sound segment whatsoever. A similar development is observed for Aleut reflexes of PAU **l*, **y*, and **h* (see table 2.7). All four of these consonant (C) segments are non-plosive, non-strident, and non-nasal. The production of plosive sounds or stops involves the stoppage of air such as is observed in the production of *p, t, k, q*, and *ʔ* (stops or plosives), that of strident sounds involves a constriction of air movement such as is observed for *č* and *s*, and that of nasal sounds involves forcing sonic air through the nasal cavity such as is observed for *m* and *n* (Burt 1974). Given this information, we can set up a single context-sensitive rule for Aleut reflexes of certain PAU consonants:

PAU C_1 → Aleut *Ø* (nil) / #__
Where C_1 is non-plosive *and* non-strident *and* non-nasal (**w*, **l*, **y*, and **h*)

Table 2.7 Consonant and Vowel Correspondences Across Aleut (A), Proto-Utian (PU), Proto-Miwok (PMi), and Proto-Costanoan (PCo) Languages, with Reconstructions of PAU Sound Segments from Which the Latter Developed (left column), and with ID Numbers (from 2.4.6) of Supporting Comparative Sets (right column)

<table>
<tr><th colspan="6">Consonant Correspondences</th></tr>
<tr><th>PAU</th><th>A</th><th>PU</th><th>PMi</th><th>PCo</th><th>Supporting Comparative | Sets</th></tr>
<tr><td>p</td><td>h < p</td><td>p</td><td>p</td><td>p</td><td>|1,|2,|3,|4,|5,|6,|</td></tr>
<tr><td>m</td><td>m</td><td>m</td><td>m</td><td>m</td><td>|17,|59|64,|65,|</td></tr>
<tr><td>w</td><td>#Ø, −y</td><td>w</td><td>w</td><td>w</td><td>|7,|8,|16,|19,|34,|44,|62,|</td></tr>
<tr><td>t</td><td>t</td><td>l</td><td>l</td><td>r</td><td>|13,|20,|29,|46,|51a,|</td></tr>
<tr><td>ṭ</td><td>t</td><td>ṭ</td><td>ṭ</td><td>ṭ, t</td><td>|9,|10,|11,|12,|13,|14,|24,|35,|65,|66,|</td></tr>
<tr><td>č</td><td>č</td><td>č</td><td>č</td><td>č, tʸ</td><td>|4,|15b,|16,|17,|18,|28,|39,|60,|61,|</td></tr>
<tr><td>s</td><td>s</td><td>s, č</td><td>s, č</td><td>s</td><td>|19,|20,|62,|</td></tr>
<tr><td>ṣ</td><td>s</td><td>ṣ</td><td>ṣ</td><td>ṣ, s</td><td>|21b,|22,|40,|</td></tr>
<tr><td>n</td><td>n</td><td>l</td><td>l</td><td>n</td><td>|6,|8,|23,|27,|36,|41,|</td></tr>
<tr><td>l</td><td>#Ø, −l</td><td>l</td><td>l</td><td>l, r</td><td>|2,|5,|7,|18,|22,|23,|24,|25,|26,|37,|42,|45, 50a,|52,|53,|54,|55,|58,|59,|65,|</td></tr>
<tr><td>k</td><td>k</td><td>k</td><td>k</td><td>k</td><td>|14,|21b,|25,|26,|27,|28,|29,|43,|</td></tr>
<tr><td>q</td><td>q</td><td>k</td><td>k</td><td></td><td>|30,|31,|32,|33,|34,|35,|36,|37,|38,|39,|48, |60,|63,|</td></tr>
<tr><td>ʁ</td><td>ʁ</td><td>q</td><td>k < q</td><td></td><td>|50a,|53,| (We recommend only rare use of fewer than three sets per correspondence in traditional evaluation [cf., 1.2.1.4].)</td></tr>
<tr><td>y</td><td>#Ø, −y</td><td>y</td><td>y</td><td>#r, −l, −y</td><td>|10,|15b,|40,|41,|42,|47,|49,|54,|56,|</td></tr>
<tr><td>ʔ</td><td>Ø</td><td>ʔ</td><td>ʔ</td><td>ʔ</td><td>|43,|44,|45,|46,|47,|48,|49,|50a,|51a,|52,|53,|54,|</td></tr>
<tr><td>h</td><td>Ø</td><td>h</td><td>h</td><td>h</td><td>|55,|56,|57,|58,|</td></tr>
</table>

(Continued)

Table 2.7 (Continued)

Vowel Correspondences

<table>
<tr><th>PAU</th><th>A</th><th>PU</th><th>PMi</th><th>PCo</th><th>Supporting Sets</th></tr>
<tr><td>i</td><td>i</td><td>#Ø, −i</td><td>#Ø, −i</td><td>#Ø, −i</td><td>|23,|59,|60,|61,|62,|62,|63,|</td></tr>
<tr><td>ɪ</td><td>a</td><td>i</td><td>i</td><td>i</td><td>|1,|9,|10,|19,|24,|28,|30,|30,|66,|</td></tr>
<tr><td>e</td><td>i</td><td>e</td><td>e</td><td>e, i</td><td>|7,|23,|43,|52,|</td></tr>
<tr><td>æ</td><td>a</td><td>e</td><td>e</td><td>e</td><td>|26,|37,|40,|44,</td></tr>
<tr><td>ɨ</td><td>u</td><td>e</td><td>e</td><td>e</td><td>|2,|27,|45,|55,|</td></tr>
<tr><td>a</td><td>a</td><td>#Ø, −a</td><td>#Ø, −a</td><td>#Ø, −a, −e</td><td>|3,|5,|8,|10,|15b,|16,|16,|20,|21b,|26,|27, |31,|32,|33,|34,|38,|45,| 46,|47,|49,|50a, |51a,|54,|55,|56,|57,|60,|64, |65,|66,|</td></tr>
<tr><td>u</td><td>u</td><td>u</td><td>u</td><td>u</td><td>|4,|5,|15b,|21b,|28,|48,|56,|</td></tr>
<tr><td>ʊ</td><td>a</td><td>u</td><td>u</td><td>u</td><td>|11,|14,|24,|25,|35,|36,|41,|41,|43,|48,|49,|57,|61,|64,|</td></tr>
<tr><td>o</td><td>u</td><td>o</td><td>o</td><td>a, o, u</td><td>|12,|13,|22,|22,|37,|42,|50a,|51a,|62,|63|</td></tr>
<tr><td>ɔ</td><td>a</td><td>o</td><td>o</td><td>o, a</td><td>|6,|14,|17,|18,|29,|38,|52,|53,|53,|54,|58,|66,|</td></tr>
</table>

Source: Generated by the authors.

Aleut shows two exceptions to this generality. First, Aleut *h* is not expected to occur in the stem-initial position; nevertheless, it is found so positioned in Aleut words in six comparative sets (1–6). In each of these situations, Aleut *h* has developed from Aleut **p* which is itself a reflex of PAU **p*. Aleut stem-initial *h* from **p* is a late development in the language, occurring after the above rule had run its course. Consequently, this *h* was not deleted in Aleut stem-initial position.

The stem-initial position in Proto-Utian involves a conditioned change. The language's reflexes of PAU **i* and **a* respectively are both realized as *Ø* (nil) stem initially, but elsewhere as expected vowels.

One other conditioned change is observed for Aleut. PAU **o* is realized as *Ø* (nil) in Aleut when the vowel is immediately preceded by **m* and immediately followed by **l* (PAU **-mol-* → Aleut *-ml-*). This change is supported by only two comparative sets, 59 and 65, and thus could be a chance development. If not, the accumulation of additional comparative sets may eventually document PAU **-mVl-* → Aleut *-ml-* with the participation of vowels other than just **o*.

Few context-conditioned phonological changes are observed for PAU sound segments, almost certainly because of the great time depth of PAU and the consequential substantial attrition of vocabulary over time.

2.4.4 Proto-Aleut-Utian Phonotactics

PAU consonants are found only stem initially: *p, ʔ, h*
PAU consonants never found stem initially: *m, n, r, ʁ*
PAU consonants found in all stem positions: *w, t, ṭ, č, s, ṣ, l, k, q, y*

Canonical shapes of PAU stems and their frequencies among the sixty-six cognate sets (C = consonant, V = vowel):

CV = 8
CVC = 17
CVCV = 30
CVCVC = 3
VCV = 1
VCVC = 3
VCVCV = 3
VCVCVC = 1

PAU words with the canonical shapes CVC and CVCV are exceptionally common, indicating these to be the most typical shapes for non-inflected bases of the reconstructed lexicon.

2.4.5 Vowel Assimilation

Bergsland (1994:xxxii) notes "several sporadic cases" of vowel assimilation in Aleut, including both progressive and regressive types. Recalling Aleut has only a three-vowel system, *i*, *a*, *u*, examples he cites of assimilation include *a* – *a* from *u* – *a*; *i* – *i* from *a* – *i*; and *u* – *u* from *u* – *i* and *u* – *a*, some of these being *tana-* "land," Proto-Eskimo (PE) *nuna* "land"; *čikiiDa* "wife's brother," PE *caki* "parent-in-law"; *chugu* "sand," Proto-Inuit *ciguraq* "sand"; *uyu-* "neck," PE *uya(qur)* "neck." (Although not stated as such, Bergsland assumes the latter pairs of Aleut-Eskimo lexemes to be cognates; in our interpretation, they are loanwords from Aleut into Eskimo or vice versa [5.5].)

In our search for Aleut-Utian cognates, instances of vowel assimilation appear to be more common than just sporadic. Among the sixty-six Aleut-Utian comparative sets (2.4.6), twenty-three or 35 percent (23/66) involve some form of vowel assimilation, for example, 10a A *tayaʁu-χ* "man," PU[a] *ṭa(:)yi-ṣ* "man" (progressive assimilation in Aleut); and 57a A *ama* "fish eggs," PU[a] **hu:ma* "fish eggs" (regressive assimilation in Aleut). (Numbers preceding examples are for identification.) Vowel assimilation involving PU is also common: 15b A *čuya-χ* "willow twig, white willow," PU[a] *čaya* "willow" (regressive assimilation for Proto-Utian); 21b A *suka-χ* "baidarka skirt," PU[a] **ṣak(:)as* "boat" (regressive assimilation for Proto-Utian).

The lower-case letter, a or b, sometimes found to be suffixed to the ID number, can indicate respectively assimilation for the Aleut word and that for the PU word. Most comparative sets of our list of sixty-six show ID numbers lacking a suffix. The comparative sets identified by the latter do not involve vowel assimilation, for example,

29. PAU **kVtɔ* FOOT (where V indicates vowel of unknown quality)
A *kita-χ* "foot (of person, bird, reindeer)"
PU **kolo* "foot"

In some instances, a suffixed ID number is the same as a non-suffixed ID number. For example,

29b. PAU **kitɔ* FOOT
A *kita-χ* "foot"
Early PU[a] **kilo* "foot" → PU **kolo* "foot" (regressive assimilation)

These two comparative sets are the same, except set 29 does not entail vowel assimilation while 29b does (regressive assimilation for Proto-Utian, where the *i* of the Early PU word is replaced by *o* in the Later PU word).

In the list of sixty-six comparative sets, where two sets show the same ID number and one is suffixed (with a or b) and the other is not, sets showing the same ID number are counted only once in the grand total of 66. This raises the question "Which of the two sets with the same ID number supports the more accurate reconstruction for the PAU "foot" word, 29 with **kVtɔ*, or 29b with PAU **kitɔ*?" Among the sixty-six sets, eighteen involve pairs of sets supporting competing PAU lexical reconstructions. There may be both positive and negative reasons to decide on one or the other competing set; however, a prolonged discussion of such complications must wait for future research.

Among the sixty-six sets, four have suffixed ID numbers showing no competing set with the same non-suffixed ID number: 15b, 21b, 50a, and 51a (see 2.4.6). These sets, each of which entails vowel assimilation, have only one PAU lexical reconstruction each.

A statistical breakdown of types of possible assimilation found among the sixty-six sets may provide some insight for distinguishing specious from real assimilation. (The following does not include comparative set 66 due to its ambiguity with respect to assimilation.)

Total consonant assimilation: 0
Total vowel assimilation: 21
Total regressive assimilation: 15
Total progressive assimilation: 6
Total Aleut assimilation: 14
Total Utian assimilation: 7
Total Aleut regressive assimilation: 8
Total Utian regressive assimilation: 7
Total Aleut progressive assimilation: 6
Total Utian progressive assimilation: 0

Based on these statistics, Aleut-Utian assimilation consists of these features: assimilation pertaining only to vowels (not to consonants), regressive assimilation pertaining to both Aleut and Utian, and progressive assimilation pertaining only to Aleut.

2.4.6 Aleut/Utian Comparative Sets

The following are the sixty-six comparative (cognate) sets for Aleut-Utian languages with associated PAU lexical reconstructions:

1. PAU **pɪ-* TO FASTEN
A *hačit-*, *hačisix* "to fasten" (derived from Early Aleut **pa-*)

PU[a] *piṭ-* "tighten"
PMi *piṭ-* "tighten"
Mins *piṭ-ki-* "to snap, clamp, fasten"

2. PAU **pɨl-* BLINK (or **p'ɨl-* [6.7])
A *huluχ-six* "to blink, wink (of eyes)" (derived from Early Aleut **puluχ* or **p'uluχ*)
PU **pel-* "pertaining to the eyes" (derived from Early PU[a] **p'el-* [6.7])
PMi[a] **pel-ak* "to blink"
Mil *pel-ak* "to blink"
PCo[a] **pel-ṭe* "to close one's eyes"
Csjb *pel-ṭe* "to close one's eyes"

Same set but with vowel assimilation posited for Aleut:

2a. PAU **pɨl-* BLINK
EA **pel-uχ* → EA **puluχ* (regressive assimilation → A *huluχ*
PU **pel-*

3. PAU **pa-* OTTER
A *halu(−χ)* "river otter" (Derived from Early Aleut **palu(−χ)*)
PU[a] **paw* "sea otter"
PMi[a] **paw* "sea otter"
Mim *paw* "otter"
Coast Miwok [*paw*] "sea otter"

4. PAU **pučV-* TURN OVER
A *huča-* "to tip over, turn over" (Derived from Early Aleut **puča*)
PU[a] **puč(:)el-* "turn over"
PMi[a] **puč(:)el-* "turn over"
PMiw **puč(:)el* "to roll around, roll over"
Mib *puc:el* "turn over"

5. PAU **pul(y)a-* BERRY TYPE (possibly, **p'ul(y)a-* [6.7])
A *hulaχ* "bearberry, *Arctostaphylos uva-ursi*" (Derived from Early Aleut **pulaχ* or **p'ulaχ*)
PU[a] **pulya* "toyon berry" (Derived from Early PU[a] **p'ulya-* [6.7])
PMi[a] **pulya* "toyon berry"
Mib *pulya* "toyon berry" (*Heteromeles arbutifolia*)

6. PAU **pɔnV-* LAKE, POND, POOL
A *hani-χ* "lake" (Note: Derived from Early Aleut **pani-*)

PU[a] **pol-* "pond, lake"
PMi **pol-ku* "to flood something"
PMis **pol:ay-* "pond, puddle, pool"
Mil *polpol* "lake"

7. PAU **wel-* SEARCH FOR, LOOK FOR
A *ilɣa-lix* "to search, to look for"
PU **wel-* "to look for, seek"
PMi[a] **wel-ṣɨ* "to look for something"
PMis **wel-ṣɨ* "to look for something"
PCo[a] **welye* "to look for something"
Ceb *werye* "to look for something"

8. PAU **wan-* SPIRIT
A *anʁ-* "spirit, spiritual being . . . to breathe" (Derived from Early Aleut **anəʁ*)
PU[a] **wali* "spirit" (PU **wal:i(p)* "earth, world" [Callaghan 2014:433])
PMi[a] **wali* "spirit"
Mib *wali* "spirit" (Collier and Thalman 1991:112)

9. PAU **ṭɪ-* FOREHEAD
A *tani-χ* "forehead"
PU[a] **ṭik:i(t)* "forehead"
PMi **ṭik:i(t)* "forehead"

10. PAU **ṭayɪ-* MAN
A *tayaʁu-χ* "man, male human being"
PU[a] **ṭa(:)yi-ṣ* "man (male)"
PCo **ṭa(:)li-ṣ* "man (male)"
Mim *tayiṣ* "man (male)"
Mib *tayih* "man (male)"

Same set but with vowel assimilation posited for Aleut:

10a. PAU **ṭayi-* MAN
EA **tayiʁu-χ* → A *tayaʁu-χ* (progressive assimilation)
PU[a] **ṭa(:)yi-ṣ*

11. PAU **ṭu-* SHINE, TO POLISH
A *takit-* "to polish, to varnish"
PU[a] **ṭul-* "shine"
PMi **ṭul-* "shine" (R: Miss *ṭul-ya-* "to polish")

12. PAU **ṭo-* SPOT, DOT, SPOTTED, DOTTED
A *tuχi-l* "to be dotted," *tuɣi-χ* "spot, dot"
PU[a] **ṭona-* "spotted, speckled"
PMi[a] *ṭona-* "spotted, speckled"
PMie **ṭona-ṭ-:a-* "spotted, speckled"

13. PAU **ṭot-* HEAR (or **ṭot'-* [6.7])
A *tut-* "to hear, get to hear" (Derived from Early Aleut[a] **tut'-* [6.7])
PU **ṭole-* "hear" (Derived from Early PU[a] **ṭol'e-* [6.7])
PMi[a] **ṭole:-yɨ* "to hear"
PMis **ṭole:-yɨ* "to hear, understand"
PCo **ta:he* "to hear"

14. PAU **ṭɔkʊ* LOOSEN
A *taka-lix* "o come off, to get loose, to go away," *takat-* "to take off, tear off, take loose (e.g., seams) . . . to unlock"
PU[a] **ṭok:u-* "to loosen, untie"
PMi[a] **ṭok:u-* "to loosen, untie"
PMis **ṭok:u-* "to loosen, untie"

Same set but with vowel assimilation posited for Aleut:

14a. PAU **ṭɔku-* LOOSEN
EA **taku-lix* → A *taka-lix* (progressive assimilation)
PU[a] **ṭok:u-*

15b. PAU **čuya* WILLOW
A *čuya-χ* "willow twig, white willow"
Early PU[a] **čuya* "willow" → PU[a] *čaya* "willow" (regressive assimilation)
PMi[a] **čaya* "willow"
Mim [*tcaiya*] "willow . . . arroyo willow . . . *Salix lasiolepis*"

16. PAU **čawaC-* FISH TRAP
A *čaya-χ* "fish pot in trap, weir, fish trap in river"
PU[a] **čáwah* "fish net"
PMi[a] **čáwah* "fish net"
Mil *čáwah* "fish net"

17. PAU **čɔm-* WASH
A *čamʁu-lix* "to wash"
PU[a] **čom:a* "to rinse, wash"
PMi **čom:a* "to rinse, soak"
Mip *com:a-* "to rinse or wash with a plunger"

18. PAU **čɔ(qa)l-* SEAWEED, KELP
A *čaqalka-n* "kind of seaweed"
PU[a] **čol:a* "kelp, seaweed"
PMi[a] **čol:a* "kelp, seaweed"
Mib *čol:a* "kelp"
Mim *cola* "seaweed"

Same set but with vowel assimilation posited for Aleut:

18a. PAU **čo(qa)l-* SEAWEED, KELP
EA **čuqalka-n* → A **čaqalka-n* (regressive assimilation)
PU[a] **čol:a*

19. PAU **sɪw-* SHELLFISH SPECIES (*Mopalia spectabilis* or *M. hindsii*?)
A *sayɣi-χ* "mossy chiton, *Mopalia muscosa*" (More likely, *Mopalia spectabilis* or *M. hindsii* since *M. muscosa* occurs in a range considerably removed from the Aleutian Islands.)
PU[a] **siwentak* "shellfish"
PMi[a] **siwentak* "shellfish"
Mim [*siwüntak*] "shellfish"

20. PAU **sat-* MUSSEL
A *satmaayu-n* "big blue mussels"
PU[a] **salo* "mussel"
PCo[a] **salo* "mussel"
Cscr [*saro*] "mussel"

21b. PAU **ṣuka-* BOAT
A *suka-χ* "baidarka skirt"
Early PU[a] **ṣuk(:)as* "boat" → Later PU[a] **ṣak(:)as* "boat" (regressive assimilation)
PMi **ṣak(:)as* "boat"
Mib *ṣakah* "boat"

22. PAU **ṣolo-* ROAR OF OCEAN
A *sulux-six* "to thunder, to make great noise, to roar (of ocean)"
PU **ṣolo* ~ **čolo* "to roar (river)" (Callaghan 2014:52–53)
PU[a] **čolo:-tu-* "flow, gush"
PMie **čolo:*-tu- "flow, gush"
PCo[a] **solo* "to flow"
Csjb *soro* "to flow"

23. PAU **line* AIR, SKY
A *ini-χ* "sky, heaven, air"
PU[a] **li:le* "air, sky"
PMi **li:le* "up, above"
Mim *lile* "up, air, sky"
PCo **ri:ni(h)* "above"

24. PAU **luṭɪ*- ONE
A *ataqa*- "one, only one, alone"
PU[a] **luṭ:i*- "one"
PMi[a] * *luṭ:i*- "one"
PMie **luṭ:i*- "single, one"

Same set but with vowel assimilation posited for Aleut:

24a. PAU **luṭi-*
EA *utiqa*- → *A ataqa*- (two successive applications of regressive assimilation [right to left])
PU[a] **luṭ:i-*

25. PAU **lʊka*- FISH LINE
A *akaɣðaχ* "fish line"
PU[a] **luk* "fish line"
PCo[a] **luk* "fish line"
Cscr *luk* "fish line"

Same set but with vowel assimilation posited for Aleut:

25a. PAU **luk*a- FISH LINE
EA *ukaɣðaχ* → A *akaɣðaχ* (regressive assimilation)
PU[a] **luk*

26. PAU **ka(m)læ*- WAVE(S) (see also comparison 45)
A *kamla-l* "to swell, to forms swells or large waves"
PU[a] **kalen* "wave"
PCo[a] **kalen* "wave"
Csol *or* Cscr [*ka-lĕn'*] "wave"

27. PAU **kanɨ* - SNOT, PHLEGM
A *kanu-n* "nasal mucus, snivel"
PU[a] **ka:leṣ* "spittle, phlegm"
PMi[a] **ka:leṣ* "spittle, phlegm"
PMie **ka:leṣ* "spittle, phlegm"

28. PAU **-kučɪ-* SMALL (or **-k'učɪ-* [6.7])
A *-kuča-* "small, small-sized" (Form is a suffix, possibly derived from Early Aleut **-k'učɪ-*)
PU **kuč:i* "small, nice"
PMi[a] **kuč:i* "small"
Mil *k'uč:i* "small"
Mic *kuci-* "good"
PCo[a] **ku:tʸi-* "small"
Csjb *ku:tʸi-* "small"

29. PAU **kVtɔ* FOOT
A *kita-χ* "foot (of person, bird, reindeer)"
PU **kolo* "foot"
PMi[a] **kolo* "foot"
PMie **kolo-* "foot"
PCo **koro:ṣum* "on foot, by means of foot"

Same set but with vowel assimilation posited for Proto-Utian:

29b. PAU **kitɔ* FOOT
A *kita-χ*
Early PU[a] **kilo* → PU **kolo* (regressive assimilation)

30. PAU **qɪCɪ* RIPE, COOKED, DONE
A *qaŋa-lix* "to be done, cooked . . . to ripen (of berries . . . of corn, seeds . . .)"
PU[a] **kili* "ripe, well done"
PMi **kili* "ripe, well done"
Mics *kil-ŋe-* "to be cooked"

31. PAU **qa-* FISH, COHO or SILVER SALMON (*Oncorhynchus kisutch*)
A *qa-χ* "fish" (Note: *qam aɣaluuʁii* "silver salmon" and *qakiiða-χ* "silver salmon") (*Oncorhynchus kisutch*)
PU[a] **ka:*ṣi "salmon," "coho" (*Oncorhynchus kisutch*)
PMi[a] **ka:ṣi* "salmon," "coho" (*Oncorhynchus kisutch*)
PMiw **kaṣ:i*, **ka:ṣi* "salmon"
Mim *káaṣi* "coho" (*Oncorhynchus kisutch*)

32. PAU **qa-* KELP, SEAWEED
A *qaŋu-χ* "seaweed, kelp (*Alaria fistulosa*)," *qasaðɣi-n* "kelp with holes in it"
PU[a] **kap:ah* "type of seaweed"
PMi[a] **kap:ah* "type of seaweed"

Mib *káp:ah* “A type of brownish-green, inedible seaweed which grows in Bodega Bay. It is long, leafless, stringy, and kelp-like; about the thickness of a person’s finger. Ducks eat the tender stem, which is apparently somehow rooted in the Bay, and the leaves drift ashore.”

33. PAU **-qa-* PAST TENSE
A *-qa-* “remote tense: remote past or, with certain auxiliaries . . . , remote future”
PU^{a} **ka:* ~ **ka* “past tense marker”
PMi **ka:* ~ **ka* “past tense marker”
Mip *-ka-* “remote past”

34. PAU **qawV-* BERRY, ESPECIALLY CROWBERRY (or **qaw’V-* [6.7])
A *qaayu-χ* “berry in general, especially crowberry, *Empetrum nigrum*” (Possibly derived from Early Aleut **qaay’u-χ* [6.7])
PU^{a} **kawisu* “wild currant” (Derived from Early Proto-Utian^{a} **qaw’isu* [6.7])
PMi^{a} **kawisu* “wild currant”
Mim *kawisu* “wild currant: bush and fruit”

35. PAU **qʊṭ-* RAZOR CLAM (*Siliqua patula*)
A *qatmi-χ* “razor clam” (Probably “Pacific razor clam,” *Siliqua patula*)
PU^{a} **ku:ṭa* “Washington clam” (Derived from Early PU^{a} **qu:ṭa* [6.7])
PMi^{a} **ku:ṭa* “Washington clam”
Mim *kúuṭa* “Washington clam”
Mib *kúuṭa* “Washington clams, *Saxidomus nuttalli*”

36. PAU **qʊn-* BONE
A *qanɣa-χ* “bone, shell”
PU^{a} **kulum* “bone” (Derived from Early PU^{a} **qulum* [6.7])
PMi^{a} **kulum* “bone”
PMiw **kulum* “bone”

Same set but with vowel assimilation posited for Proto-Utian:

36b. PAU **qan-* BONE
A *qanɣa-χ*
Early PU^{a} **kalum* → PU^{a} **kulum* (regressive assimilation)

37. PAU **qolæ-* GRASS IN GENERAL, GRASS TYPE (or **qol’æ-* [6.7])
A *qulaaluχ* “sedge (*Carex sp.*) and similar kinds of grass such as brome (. . . *Bromus aleutensis*) and reed (*Calamagrostis langsdorfii*), used for weaving” (derived from Early Aleut *qul’aaluχ* [6.7])

PU[a] *kole* "grass, hay" (Derived from Early PU[a] *qol'e* [6.7])
PMi[a] **kole* "grass, hay"
PMiw **kole* "grass, hay"
Mib *kóle* "grass (generic), hay"

38. PAU **qɔCa-* SEA LION
A *qawa-χ* "sea lion, *Eumetopias jubata*" (Steller sea lion)
PU[a] kota "California sea lion"
PMi[a] **kota* "California sea lion"
Mim *kóta* "California sea lion" (*Zalophus californianus*)

Same set but with vowel assimilation posited for Aleut:

38a. PAU **qoCa* SEA LION
EA **quwa-χ* → A *qawa-χ* (regressive assimilation)
PU[a] *kota*

39. PAU **qVčV-* BONE, LEG BONE (or **q'VčV-* [6.7])
A *qačima-χ* "thighbone, femur"
PU[a] **kɨč:ɨč* "bone" (Partial reduplication of **kɨč:*?) (Derived from Early PU[a] **q'ɨč:ɨč* [6.7])
PMi[a] **kɨč:ɨč* "bone"
PMie **kɨč:ɨč* "bone"

40. PAU **yæṣ-* FIELD MOUSE (*Microtus*)
A *asχukiða-χ* "mouse" probably field mouse, "*Microtus*" ("field mouse" and "meadow vole" are both names for *Microtus pennsylvanicus*)
PU[a] **yeṣ:e* "mouse"
PMi[a] **yeṣ:e* "mouse"
PMiw **yeṣ:e* "mouse"
Mim [*yes'-se*] "*Microtus californicus*"

Same set but with vowel assimilation posited for Proto-Utian:

40b. PAU *yaṣ- FIELD MOUSE (*Microtus*)
A *asχukiða-χ*
Early PU[a] **yaṣ:e* → PU[a] **yeṣ:e* (regressive assimilation)

41. PAU **yʊnʊ-* SMALL RODENT (LEMMING, MOUSE)
A *anamču-χ* "kind of mouse . . . without tail . . . possibly lemming"
PU[a] **yul:u* "round-tail woodrat, *Neotoma*"
PMi[a] **yul:u* "round-tail woodrat, *Neotoma*"
PMiw **yul:u* "round-tail woodrat, *Neotoma*"

42. PAU **yolV-* RED
A *uluuða-χ, uluuya-χ* "red"
PU **yoli* "red"
PMi[a] **yoli-* "red, a red one"
Mics *yol-ol:i* "red, a red one"
PCo[a] **yuli-ț* "redness"
Cru *yuri-ț* "redness"

Same set but with vowel assimilation posited for Aleut:

42a. PAU **yoli* RED
EA **uliuða-χ* → A *uluuða-χ* (progressive assimilation)
PU **yoli*

43. PAU **ʔeku* HAND, FINGER
A *ikalqu-χ* "ring finger, fourth finger"
PU **ʔek:u* "hand, finger"
PMi **ʔek:u-s* "hand, finger"
PCo **ʔis:u* "hand, arm" (*s:* is not explained)

44. PAU **ʔæw-* BASKET
A *ayɣaʁasi-χ* "basket, berry basket of grass"
PU[a] **ʔéwih* "basket"
PMi[a] **ʔéwih* "basket"
Mib *ʔéwih* "basket (any type)"

45. PAU **ʔali* - WAVE(S) (See also comparison 26.)
A *aalu-χ* "swell, big wave"
PU[a] **ʔa:len* "wave(s)"
PCo[a] **ʔa:len* "waves"
Cru [*ah len*] "waves"

46. PAU **ʔaț-* UNDERSIDE
A *at-* "lower part, area below, underside, bottom"
PU[a] **ʔa:la-* "below, underside"
PMi **ʔa:la-* "below, east"
Miss *ʔal:a-* "underside"
Mics *ʔal:a-* "down, low"

47. PAU **ʔay-* MORE
A *ayxa-lix* "to exceed, be more or greater etc. than"
PU[a] **ʔa:y* "more"
PMi **ʔa:y* "again, another, more," *ʔa:y-ɨ-* "to increase"

48. PAU **ʔʊqu-* DOOR, GATE
A *aquli-χ* "gate"
PU[a] **ʔu:ku-* "door"
PMi **ʔu:ku-* "to enter, to go in, to set (sun)"
PMie **ʔuku-y:a-* "door"

Same set but with vowel assimilation posited for Proto-Utian:

48b. PAU **ʔaqu-* DOOR, GATE
A *aquli-χ* "gate"
Early PU[a] **ʔa:ku-* "door" → PU[a] *ʔu:ku-* (regressive assimilation)

49. PAU **ʔʊya-* ASHAMED, BASHFUL
A *ayaʁi-lix* "to be ashamed, to be bashful"
PU **ʔuy(:)a* "ashamed"
pMi[a] **ʔuya-* "bashful, ashamed"
PMiw **ʔoya-ṣ* "bashful, ashamed"
PCo **ʔuy:a-n* "to be ashamed"

Same set but with vowel assimilation posited for Aleut:

49a. PAU *ʔuya-* ASHAMED, BASHFUL
EA **uyaʁi-lix* → A *ayaʁi-lix* (regressive assimilation)
PU **ʔuy(:)a*

50a. PAU **ʔolaʁ-* HATCH OF BOAT, OAR, ROW
EA **ulaʁu-χ* "hatch of boat" → *A uluʁu-χ* "hatch of boat" (progressive assimilation)
PU[a] **ʔo:lak* "oar, to row" (Derived from Early PU[a] **ʔo:laq* [6.7], see 53)
PMi[a] **ʔo:lak* "oar, to row"
PMiw **ʔo(:)lak* "oar"
Mil *ʔo:lak* "to row"

51a. PAU **ʔota* SEA ANEMONE
Early A **uta-χ* "sea anemones"→ A *utu-χ* "sea anemones" (progressive assimilation)
PU[a] **ʔo:la* "sea anemone"
PMi[a] **ʔo:la* "sea anemone"
Mib *ʔóola* "sea anemone"

52. PAU **ʔɔle-* WILD CANINE (COYOTE, WOLF)
A *aliiʁi-χ* "wolf (on Alaska Peninsula and Unimak Island)" (No Aleut word for coyote.)

PU[a] *ʔole "coyote"
PMi *ʔole "coyote" (PMi *hu:n ~ hun:u "wolf".)

53. PAU *ʔɔlɔʁ- OCEAN, SEA
A *alaʁu-χ* "the sea, the ocean"
Early PU[a] *ʔo:loq* "ocean"
Later PU[a] *ʔo:lok "ocean" (Derived from Early PU[a] *ʔo:loq* [6.7], see 50a)
PMi[a] *ʔo:lok "ocean"
Mib *ʔóolok* "ocean"

54. PAU *ʔɔya- FRIEND
A *aya-* "friend, cousin"
PU[a] *ʔoy:am "friend, partner, in-law"
PMi *ʔoy:am "friend, partner, in-law"

Same set but with vowel assimilation posited for Aleut:

54a. PAU *ʔoya- FRIEND
EA *uya- → A *aya-* (regressive assimilation)
PU[a] *ʔoy:am

55. PAU *(-)hɨla NO, NOT
A *-ulax* "enclitic negative"
PU[a] *hel:a "no, not"
PMi *hel:a "no, not"

56. PAU *hay(k)u DOG
A *ayku-χ* "dog"
PU[a] *hayu "dog"
PMi *hayu "dog"

57. PAU *hʊma FISH EGGS (ROE)
A *ama-χ* "fish roe . . . vulva"
PU[a] *hu:ma "salmon eggs, fish eggs"
PMi[a] *hu:ma "salmon eggs, fish eggs"
Mib *húuma* "salmon eggs, fish eggs"

Same set but with vowel assimilation posited for Aleut:

57a. PAU *huma FISH EGGS (ROE)
EA *uma → A *ama* (regressive assimilation)
PU[a] *hu:ma

58. PAU **hɔl-* BUTTERFLY
A *almiχta-χ* "butterfly"
PU[a] **hol(:)o* "butterfly"
PMi[a] **hol(:)o* "butterfly"
PMie **hol(:)o . . .* "butterfly"

Same set but with vowel assimilation posited for Proto-Utian:

58b. PAU **hal-* BUTTERFLY
A *almiχta-χ*
Early PU[a] **hal(:)o* → PU[a] **hol(:)o* (regressive assimilation)

59. PAU **imolV* HAIR OF HEAD
A *imli-χ* "single hair of head . . . hair of head" (see comparison 65 for Aleut *-ml-* cluster)
PU[a] **molu* "head, hair, body hair"
PMi[a] **molu* "head, hair, body hair"
Mib *molu* "head, hair, body hair (but not dog's hair)"

60. PAU **ičaq-* THIN
A *ičaaqiða-lix* "to be thin (as of skin, booklet)"
PU[a] **čak-* "thin"
PMi[a] **čak-* "thin"
PMis **čak-* "thin, transparent"

61. PAU **ičʊ-* FILL
A *iča-lix* "to fill"
PU[a] **ču:yaʔ-nɨ-* "to fill"
PMi[a] **ču:yaʔ-nɨ-* "to fill"
PMis **ču:yaʔ-nɨ-* "to fill"

62. PAU **isowi(q)* HARBOR SEAL (6.7)
A *isuχ* "seal, harbor seal, *Phoca vitulina richardii*" (Derived from Early Aleut **isuyiχ*)
Early PU[a] **so:-wiq* "harbor seal" (6.7)
Later PU[a] **so:-wik* "harbor seal"
PCo[a] *so:-wik "harbor seal"
Cru [*soó-wík*] "hair seal" (This is also a name for harbor seal.)
(Note: This word is also found in Yupik Eskimo languages neighboring Aleut: Kodiak Alutiiq with *isuwiq* "seal" and Central Alaskan Yupik with *issuʁiq* "spotted seal." Bergsland [1994:214] proposes the term was borrowed by these languages from Aleut. If so, the Early Aleut word was originally

something like **isuyiq*, ultimately reduced to *isuχ* in the modern language. This accords with our reasoning for reconstructing the second syllable of the PAU word [**isowiq*].)

63. PAU **iqoṭ-* BREAK
A *iqut-* "to break"
PU[a] **koṭ:a* "to break"
PMi **koṭ:a* "to break something, to break something in two" (R: Mics *koṭ-ku-* "to break a limb")

64. PAU **amʊC-* BLACK, NIGHT
A *amax* "night"
PU **mulu-* "darkness, night"
PMi **mulu-* "black, dark blue, dark"
PCo **muru-t* "darkness, night"

Same set but with vowel assimilation posited for Aleut:

64a. PAU **amuC-* BLACK, NIGHT
EA **amux* "night" → A *amax* (progressive assimilation)
PU **mulu-*

65. PAU **am(o)lu-* TO COVER
A *amlux-six* "to wrap up, cover (obsolete)" (See comparison 59 for Aleut *-ml-* cluster.)
PU **mol:ɨ* "to cover something"
PMi[a] **mol:ɨ* "to cover something"
PMiw **mol:u-* "to cover something"
PCo[a] **mali:* "to cover"
Csjb *malii* "to cover the genitals"

66. PAU **aṭɔCɪ* TO THINK
A *ataaða-lix* "to think," *ataasa-lix* "to think to be, to take for"
PU[a] **ṭon:i* "to think"
PMi[a] **ṭon:i-* "to think"
PMie **ṭon:i* "to think, to think about something"

Same set but with vowel assimilation posited for Aleut:

66a. PAU **aṭoCɪ-* TO THINK
EA **atuða-lix* → A *ataaða-lix* (progressive or regressive assimilation)
PU[a] **ṭon:i*

2.5 QUANTITATIVE VS. TRADITIONAL

Individual comparativists rarely include *both* quantitative and traditional approaches in their methodologies, most likely due to the recency of automation against a background of 200 years or so during which the traditional method has prevailed. Whatever the reason for others' choices, we view quantitative and traditional approaches as complementary, each having its own special attributes for contributing to the great goal of determining relationships among all the world's languages as far back in time as can possibly be accomplished.

We suggest quantitative evaluation (through screening combined with ASJP similarity) as a first step in undertaking comparative linguistics, determining language connections in a rigorous and systematic manner. These connections might eventually be incorporated into a large catalog, something along the lines of online appendix A (see table 1.1 for a reduced-size version), which could be consulted for the feasibility of different language relationships proposed in the past or for future study. Language comparisons receiving *Weak* or *No-Support* evaluations could be judged out of consideration before scholars investigate them in detail, saving time and other resources. Language comparisons cataloged as *Strong* might also be set aside if the goal of research were merely to prove a language affiliation, augmenting the amount of time in which the researcher considers other details, such as phonological and lexical reconstruction, using the traditional method. *Moderate* comparisons should be of greatest interest to researchers, since this result could direct them to needed additional study using the special tools of the traditional approach, as we have done for Aleut/Utian in this chapter.

Our quantitative approach has a feature capable of advancing comparative linguistics beyond just determining whether two languages are similar by chance: the incorporation of basic vocabulary into the algorithm, allowing it to distinguish lexical similarity in the compared languages due to borrowing from that due to kinship. Achieving this goal in a systematic manner has been one of the great challenges of modern historical linguistics. We believe we have come close to solving the problem and eagerly await critical responses.

2.6 CONCLUSION

This chapter uses both quantitative and traditional methods for evaluating the hypothesis that Aleut and Utian are genealogically related. While results of the quantitative approach for some comparisons are only moderately supportive, findings are nonetheless deemed worthy of additional research. Following this recommendation, the use of the traditional method yields the reconstruction of a PAU phonological inventory and a vocabulary of sixty-six words. In subsequent chapters, especially chapter 9, we investigate how these restorations contribute to the knowledge of the cultural histories of Aleut and Utian peoples.

Chapter 3

Structural Similarity of Aleut and Utian

3.1 INTRODUCTION

The preceding two chapters assemble lexical evidence for the genealogical relationship of Aleut and Utian languages. By "lexical evidence," we mean similarities pertaining to *words* of languages; the more similar the words of compared languages, the more likely these languages descend from a common ancestor. While both quantitative and traditional comparative linguistics focus on lexical similarity (chapters 1 and 2), language structural features such as phonology, morphology, and syntax are also investigated (cf. Hübler and Greenhill 2023). Some linguists have even proposed structural resemblance to be more probative of language relatedness than lexical similarity (for examples, see Campbell and Poser 2008). If this were so, features observed to be more resistant to change over time would be structural rather than lexical. To our knowledge, however, this proposal has not been empirically demonstrated or even investigated (but see 10.4.1).

Structure can be especially indicative of genealogical relationships if compared languages have in common one or more systematic association(s) of morphemes constituting a "grammatical paradigm." An oft-cited example is English *good/better/best* and German *gut/besser/best*, the Germanic suppletive positive pattern (cf. Campbell and Poser 2008:182, Nichols 2010:304). Nichols estimates a two-in-one-hundred-million probability that two unrelated languages could have such a nuanced pattern in common by chance. Any two languages sharing a structural paradigm constitute strong candidates for genealogical affiliation. We have yet to observe such a paradigm common to Aleut and Utian, but, according to Nichols (2010:304), such a negative result is expected in long-range comparisons such as ours.

This chapter examines structural evidence supporting the affiliation of Aleut and Utian. The approach used is quantitative. Specific structural features treated are gleaned from the online version of the *World Atlas of Language Structures* (*WALS*) (Dryer and Haspelmath 2013). *WALS* provides information on 192 structural features for 2,679 languages and dialects, with each feature showing from two to nine values, for example, feature 83A "Order of Object and Verb," shows three values: "OV," "VO," and "no dominant order" (where O = object, and V = verb). Values, given as circles distinguished by different colors, are charted on *WALS* world maps for individual features.

3.2 *WALS* STRUCTURAL FEATURES AND VALUES FOR ALEUT AND UTIAN LANGUAGES

Two language entries represent Aleut in *WALS*: "Aleut" with values for twenty-five features, and "Aleut (Eastern)" with values for fourteen features, none of which overlap for either entry. Information on eight Utian languages is given in *WALS*, six of which belong to the Miwok branch of the family and two to the Costanoan branch. These languages are as follows, with the number of features for each in parentheses: Bodega Miwok (2), Central Sierra Miwok (7), Lake Miwok (2), Northern Sierra Miwok (8), Plains Miwok (2), Southern Sierra Miwok (100), Mutsun (Costanoan with 37), and Rumsen (Costanoan with 18). Southern Sierra Miwok is the only Utian language referenced in this investigation because of its large number of *WALS* features (100).

All *WALS* languages with one hundred or more features (157 languages) are included in the study. Also included is Aleut, even though it has fewer than one hundred features. (Features from the two Aleut entries, "Aleut" and "Aleut [Eastern]," are combined, yielding thirty-nine features for the language.) The total number of languages used in our *WALS* study (including Aleut) is 158.

Table 3.1 lists the thirty-nine Aleut structural features and indicates values of these features also reported for Southern Sierra Miwok. Presented from left to right in the table are the *WALS* feature-identification number, the name of the feature, the value of the feature pertaining to Aleut, and, in the last column, indication of whether the value is found for the feature in Southern Sierra Miwok ("same" or "different"). Six of the thirty-nine features are not available in *WALS* for the latter language; thus, values for only thirty-three features are reported for both languages. Aleut and Southern Sierra Miwok have nineteen values in common for these thirty-three structural features, rendering a shared-value percentage (SVP) of 57.6 (19/33 × 100).

Table 3.1 Thirty-Nine *WALS* Structural Features and Their Values Reported for Aleut, Southern Sierra Miwok, and Both (If the Same, We Assume the Value Reconstructs to Proto-Aleut-Utian).

WALS NO.	*WALS Features Reported for Both Aleut and Southern Sierra Miwok*	*Aleut Value*	*Southern Sierra Miwok Value*
1A	Consonant inventories	Average	Different
2A	Vowel quality inventories	Small (2–4)	Different
3A	Consonant vowel ratio	High	Different
4A	Voicing in plosives and fricatives	In fricatives alone	Different
5A	Voicing and gaps in plosive systems	Other	Same
6A	Uvular consonants	Uvular stops and continuants	Different
7A	Glottalized consonants	No glottalized consonants	Same
8A	Lateral consonants	/l/, no obstruent laterals	Same
9A	The velar nasal	Initial velar nasal	Different
11A	Vowel nasalization	None	Same
12A	Syllable structure	Complex	Different
13A	Tone	No tones	Same
18A	Absence of common consonants	All present	Same
19A	Presence of uncommon consonants	"Th" sounds	Different
26A	Prefixing vs. suffixing in inflectional morphology	Strongly suffixing	Same
30A	Number of genders	None	Same
31A	Sex-based and non-sex-based gender systems	No gender	Same
32A	Systems of gender assignment	No gender	Same
33A	Coding of nominal plurality	Plural suffix	Same
35A	Plurality in independent personal pronouns	Person–number affixes	Different
43A	Third person pronouns and demonstratives	Related for all demonstratives	Same
49A	Number of cases	2 cases	Different
50A	Asymmetrical case-marking	Subtractive-quantitatively asymmetrical	Different
51A	Position of case affixes	Case suffixes	Feature not available
52A	Comitatives and instrumentals	Mixed	Feature not available
53A	Ordinal numerals	First, two-th, three-th	Different
54A	Distributive numerals	Marked by suffix	Feature not available
57A	Position of pronominal possessive affixes	Possessive suffixes	Same

(*Continued*)

Table 3.1 (Continued)

WALS NO.	*WALS Features Reported for Both Aleut and Southern Sierra Miwok*	*Aleut Value*	*Southern Sierra Miwok Value*
63A	Noun phrase conjunction	"and" different from "with"	Feature not available
69A	Position of tense-aspect affixes	Tense-aspect suffixes	Same
72A	Imperative-hortative systems	Maximal system	Same
81A	Order of subject, object, and verb	SOV	Different
82A	Order of subject and verb	SV	Same
83A	Order of object and verb	OV	Different
105A	Ditransitive constructions: The verb "give"	Indirect-object construction	Feature not available
121A	Comparative constructions	Locational	Same
124A	"Want" complement subjects	Desiderative verbal affix	Feature not available
129A	Hand and arm	Different	Same
130A	Finger and hand	Different	Same

Source: Generated by the authors based on data obtained from *WALS*.

With the exception of Aleut, *WALS* languages cited in this study are limited to the 157 for which one hundred or more structural features are reported. (We thank Matthew Dryer and Robert Forkel for providing a complete list of *WALS* languages ranked from largest to smallest by number of features reported for each.) Table 3.2 lists the 157 languages in rank order according to the size of SVP (shared-value percentage based on the language's comparison with Aleut), from largest to smallest. In the table, from left to right, is the *WALS*-language name with indication of the number of structural features for the language in parentheses, the genealogical affiliation of the language, the hemisphere (New World = NW, Old World = OW) in which the language was/is spoken, the distance in kilometers (as the crow flies) of the language's removal from Aleut, the number of *WALS* structural features shared with Aleut (following comma), the number of values of these features shared with Aleut (preceding comma), and the SVP. (We include only the first 57 of the 157 entries in this text version of table 3.2. The full table with 157 entries can be found in online appendix B.)

With an SVP of 57.6, that of Southern Sierra Miwok is the eleventh largest of the 157 SVPs observed for languages of table 3.2 (tied with Cahuilla). Southern Sierra Miwok, then, more closely resembles Aleut structurally than 92.4 percent of the other languages in the sample. In chapters 1 and 2, we assemble lexical evidence indicating a historical connection between Aleut and Utian

Table 3.2 Structural Similarity with Aleut (SVP) Observed for Each of the 157 *WALS* Languages for Which One Hundred Or More Features Are Reported

Language (Features)	*Classification*	*Hemisphere*	*Distance from Aleut (km) As the Crow Flies*	*No. Shared with Aleut (Values, Features)*	*SVP*
Nenets (113)	*Uralic (Uralo-Siberian)*	*OW*	*5,572*	*24, 36*	*66.7*
Evenki (149)	*Tungusic (Altaic)*	*OW*	*4,235*	*25, 40*	*62.5*
West Greenlandic (142)	*Eskimo (Uralo-Siberian)*	*NW*	*5,781*	*23, 37*	*62.2*
Central Aymara (100)	Aymaran	NW	11,974	21, 44	61.8
Yidiny (115)	Pama-Nyungan	OW	9,106	19, 31	61.3
Awa Pit (131)	Barbacoan	NW	9,726	23, 38	60.5
Haida (111)	*Haida*	*NW*	*2,237*	*22, 37*	*59.5*
Khalkha (144)	*Mongolic (Altaic)*	*OW*	*5,936*	*23, 39*	*59.0*
Turkish (154)	*Turkic (Altaic)*	*OW*	*9,505*	*24, 41*	*58.5*
Ngiyambaa (136)	Pama-Nyungan	OW	10,583	21, 36	58.3
Southern Sierra Miwok (100)	Utian	NW	3,930	19, 33	57.6
Cahuilla (107)	Uto-Aztecan	NW	4,484	19, 33	57.6
Yukaghir (136)	*Yukaghir (Uralo-Siberian)*	*OW*	*2,675*	*21, 38*	*55.3*
Georgian (150)	Kartvelian	OW	8,989	22, 40	55.0
Armenian (Eastern) (129)	Indo-European	OW	9,176	18, 33	54.6
Maricopa (134)	Yuman	NW	4,710	19, 35	54.3
Yaqui (141)	Uto-Aztecan	NW	5,363	20, 37	54.1
Chukchi (137)	*Chukotko-Kamchatkan*	*OW*	*1,496*	*21, 39*	*53.9*
Kanuri (137)	Saharan	OW	12,690	21, 39	53.9
Lower Grand Valley Dani (115)	Dani	OW	8,230	17, 32	53.1
Zoque (Copainalá) (110)	Mixe-Zoquean	NW	7,370	17, 32	53.1
Nivkh (131)	*Nivkh*	*OW*	*3,360*	*19, 36*	*52.8*

(*Continued*)

Table 3.2 (Continued)

Language (Features)	*Classification*	*Hemisphere*	*Distance from Aleut (km) As the Crow Flies*	*No. Shared with Aleut (Values, Features)*	*SVP*
Burushaski (140)	Burushaski	OW	8,429	21, 40	52.5
Mapudungun (138)	Araucanian	NW	13,552	21, 40	52.5
Basque (149)	Basque	OW	9,134	21, 40	52.5
Nubian (Dongolese) (100)	Eastern Sudanic	OW	11,841	11, 21	52.4
Daga (111)	Dagan	OW	8,240	15, 29	51.7
Korean (149)	Korean	OW	5,237	19, 37	51.4
Martuthunira (136)	Pama-Nyungan	OW	11,073	20, 39	51.3
Yimas (118)	Lower Sepik-Ramu	OW	8,006	17, 34	50.0
Hindi (144)	Indo-European	OW	9,387	18, 36	50.0
Kayardild (132)	Tangkic	OW	9,420	17, 34	50.0
Comanche (109)	Uto-Aztecan	NW	5,447	18, 36	50.0
Koasati (125)	Muskogean	NW	6,396	18, 36	50.0
Hungarian (155)	*Uralic, Ugric (Uralo-Siberian)*	*OW*	*8,797*	*20, 41*	*48.8*
Shipibo-Konibo (118)	Panoan	NW	10,743	18, 37	48.7
Rama (127)	Chibchan	NW	8,454	17, 35	48.6
Navajo (107)	Na-Dene	NW	4,815	16, 33	48.5
Beja (105)	Beja	OW	11,756	15, 31	48.4
Quechua (Imbabura) (129)	Quechuan	NW	12,471	12, 25	48.0
Modern Greek (155)	Indo-European	OW	9,671	19, 40	47.5
Guaraní (137)	Tupian	NW	13,597	19, 40	47.5
Kobon (130)	Madang	OW	8,013	18, 38	47.4
Brahui (104)	Davidian	OW	9,548	15, 32	46.9
Lezgian (149)	Nakh-Daghestanian	OW	8,924	19, 41	46.3

Finnish (155)	*Uralic (Uralo-Siberian)*	*OW*	*7,105*	*19, 41*	*46.3*
Gooniyandi (128)	Bunuban	OW	10,262	17, 37	46.0
Burmese (140)	Sino-Tibetan	OW	8,650	17, 37	46.0
Barasano (112)	Tucanoan	NW	10,372	11, 24	45.8
Imonda (126)	Border	OW	8,005	16, 35	45.7
Nez Perce (103)	Sahaptian	NW	3,612	15, 33	45.5
Asmat (122)	Asmat-Kamoro	OW	8,361	15, 33	45.5
Kannada (143)	Dravidian	OW	10,480	18, 40	45.0
Canela-Krahõ (129)	Ge-Kaingang	NW	12,471	17, 38	44.7
Rapanui (135)	Austronesian	OW	10,501	17, 38	44.7
Meithei (132)	Sino-Tibetan	OW	8,439	16, 36	44.4
Ingush (113)	Nakh-Daghestanian	OW	8,839	14, 32	43.8

Source: Generated by the authors using data from *WALS*.
Notes: Italicized entries are circumarctic languages; SVP = shared-value percentage. See the full table in online appendix B for the remaining 100 comparisons in this table. These latter comparisons have SVPs ranging in value from Kunama with 43.8 to Wari' with 8.6.

(of which Southern Sierra Miwok is a member language). Two distinct lines of linguistic evidence, then, attest to the overall close resemblance of Aleut and Utian, one *lexical* and one *structural* (see table 3.2; lines of non-linguistic evidence are considered in chapters 4 [genetics] and 8 [archaeology]).

Among the thirteen languages with the largest SVPs (ranging from 66.7–55.3), seven (or 54.0 percent) are circumarctic in geographic distribution. (In the table, circumarctic languages are enclosed in boxes with extra-thick borders.) Six of these seven have been proposed as belonging to either Uralo-Siberian (Fortescue 1998, Fortescue and Vajda 2022) or to Altaic (Robbeets and Savelyev 2020) language groups: with Nenets, West Greenlandic, and Yukaghir belonging to Uralo-Siberian (Aleut is also considered Uralo-Siberian by Fortescue); Evenki, Khalkha, and Turkish to Altaic. In a now-largely-rejected interpretation, Uralic and Altaic languages are grouped together as Ural-Altaic, a classification that more or less accords with the close structural resemblance found for the six circumarctic languages discussed here.

While advocates of these various broad groupings have proposed similarities explained by genealogical affiliation, widespread contact and borrowing (direct and indirect) over long periods of time are just as likely, if not better, explanations for shared structural resemblances (5.6), cf. Georg 2017. Whatever the case, the SVP of Southern Sierra Miwok of California also indicates its participation in developments responsible for structural similarities of circumarctic languages of Altaic and Uralo-Siberian groups.

Nenets, with 66.7 shows the closest structural similarity to Aleut, a finding that accords with Fortescue's Uralo-Siberian grouping, which includes Eskimo-Aleut. If Aleut is a constituent of the former group, then Aleut's closest kin, Utian (to which Southern Sierra Miwok belongs), should also be classified as a Uralo-Siberian language (but see 5.4).

3.3 COMBINED DOUBLE MARKING

We examine two *WALS* values with distributions supporting the historical connection between Aleut and Utian (specifically Southern Sierra Miwok). These are values of features 23A and 24A, respectively, "Locus of Marking in the Clause" and "Locus of Marking in Possessive Noun Phrases" (Nichols and Bickel 2013a, b). (Aleut values of features 23A and 24A are not given in *WALS*; however, Aleut values for these features were supplied to us by Anna Berge and Michael D. Fortescue in respective personal communications.) Values of these two features and their frequencies of occurrence in languages of the *WALS* sample are given in table 3.3.

Table 3.3 Values of Features 23A and 24A and their Frequencies of Occurrence in Languages of the *WALS* Sample

Feature 23A (Total = 236)

Values	*Frequency*
Head marking	71
Dependent marking	63
Double marking	58
No marking	42
Other	2

Feature 24A (Total = 236)

Values	*Frequency*
Head marking	78
Dependent marking	98
Double marking	22
No marking	32
Other	6

Source: Generated by the authors using *WALS* data.

Both Aleut and Southern Sierra Miwok show the value "Double marking" for both features, as do fifteen other languages of the *WALS* sample. These sixteen languages, not including Aleut whose values for the two features are non-*WALS* supplied (3.3, above), are Aymara (Central), Belhare, Burushaski, Djingli, Greek (Modern), Greenlandic (West), Hua, Imonda, Jívaro, Miwok (Southern Sierra), Mangarrayi, Paiwan, Wembawemba, Yaqui, Yupik (Central), and Zoque (Copainalá). (Fortescue [pers. com.] also reports combined double marking for all Uralo-Siberian languages.) Since the total number of *WALS* languages combining any two values of the two features is 268, the expected frequency percentage for sixteen languages with combined "Double marking" is 6.7 percent (16/238 × 100). In other words, the odds that Aleut and Southern Sierra Miwok both have these same two values by chance is about seven in one hundred. If it has not developed by chance, combined double marking for both Aleut and Southern Sierra Miwok likely results from historical connection. A probability of 6.7 percent would at best be considered only marginally convincing since statistical significance is typically considered achieved by statisticians only when results are at or less than 5.0 percent.

3.4 PRONOMINAL POSSESSIVE AFFIXES

In reference to *WALS* value mapping of this feature (no. 57A), Dryer (2013) writes, "The map shows what is perhaps the clearest apparent example in this atlas of an Old World-New World split in the distribution of the two types of possessive affixes: while possessive suffixes are the primary type in the Old World, possessive prefixes are primary in the New World." Aleut and Southern Sierra Miwok are New World languages that deviate from the pattern described by Dryer since they both show the value "possessive suffix." This deviation is not particularly surprising for Aleut, since it is geographically associated more closely with New World languages (Eskimo) showing the predominant Old World value "possessive suffix" than the usual New World value. On the other hand, Southern Sierra Miwok of California unexpectedly resides in about the middle of a large distribution of New World languages showing the "possessive prefix" (see map, figure 3.1).

Distribution of these two values conforms with the migration of Proto-Utian speakers from coastal southwest Alaska to coastal California. Proto-Aleut-Utian bequeathed the value "possessive suffix" to its two daughter languages, Aleut and Proto-Utian. Speakers of Utian in California maintained the value despite finding themselves surrounded by speakers of other languages showing possessive prefixes. With the breakup of Proto-Utian, at least one of its offspring (Southern Sierra Miwok) retained possessive suffixes.

Callaghan (2014:179–181) reconstructs, although only tentatively, possessive pronominal suffixes for Proto-Utian. This value is not found in several

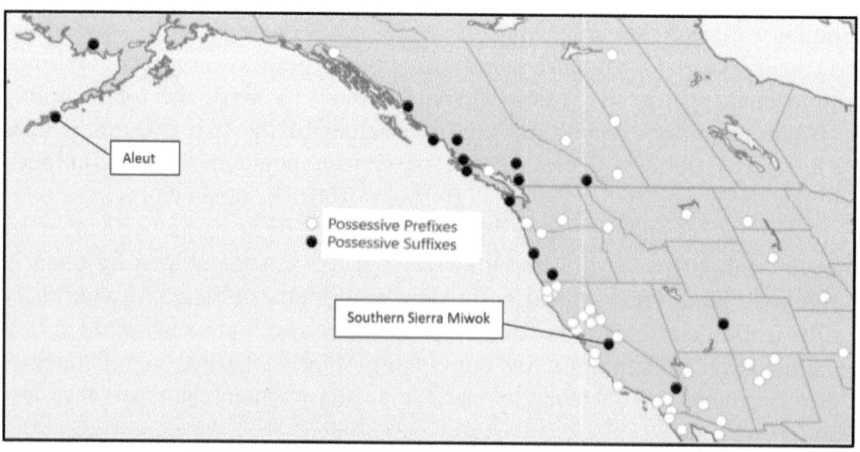

Figure 3.1 Distribution of Two Values (Designated, Respectively, by White and Black Circles) of WALS Feature 57A. *Source*: Adapted from Dryer (2013).

California Utian languages, for example, Mutsun, a Costanoan language, where the feature's realization is the "possessive prefix." Since data are limited for Costanoan languages, whether this value pertains to this subgroup in general or is only specific to Mutsun is uncertain. If found to be a value of all or most Costanoan languages, reconstruction of the "possessive suffix" for Proto-Utian would be even more tentative. Moreover, at least two Western Miwok languages, Bodega and Lake Miwok, show possessive prefixes.

After the breakup of Proto-Utian, daughter languages became scattered in distribution along the central California coast and into the interior. In their new settings, Utian speakers encountered speakers of local languages. Some Proto-Utian offspring were influenced by these languages and adopted the predominant value of their neighbors for Feature 57A, that is, the "possessive prefix." In contrast, other Proto-Utian offspring, such as Southern Sierra Miwok, continued the usage their ancestral language brought to California from Alaska, that is, possessive suffixing.

An examination of the map in figure 3.1 locates several "possessive suffix" languages (black circles) in an area of the Pacific coast extending from the Olympic Peninsula of northwest Washington State (Quileute language) northward to the central coast of British Columbia (Nuxalk language). These languages might be regarded as an extension of the value's distribution from Old World languages of East Asia to New World languages (Eskimo and Aleut) along the Pacific Rim into languages of the Northwest Coast. (Fortescue [pers. com.] informs us that the "possessive suffix" is found in all Uralo-Siberian languages except Yukaghir.) As we propose in detail in chapter 6, the journey of Proto-Utians from south Alaska to California via watercraft involved at least two prolonged layovers, respectively, in the northern and southern areas of the coastal region described above. During these stopovers, under the influence of Proto-Utian, native "possessive prefix" languages plausibly shifted to the "possessive suffix."

3.5 STRUCTURAL RECONSTRUCTIONS FOR PROTO-ALEUT-UTIAN

In chapter 2, we reconstruct a fragment of the PAU vocabulary (66 protowords). The small size of the recoverable PAU lexicon is not unexpected given the great span of time since its divergence into Aleut and Proto-Utian (sometime during the Middle Holocene, 9.2), and the occurrence of substantial loss and replacement of vocabulary since that time. However, the reconstruction of a few PAU structural features is possible, and these are likely the same nineteen features and their values as designated in table 3.1 by "Same" for Southern Sierra Miwok.

3.6 CONCLUSION

Southern Sierra Miwok more closely resembles Aleut structurally than do 92.4 percent of the other 156 languages of our *WALS*-language sample. These two languages also have in common nineteen values out of a possible thirty-three (table 3.1). While these findings may not be considered incontrovertible evidence for the historical connection of Utian and Aleut, they do not controvert the proposal that Aleut and Utian share a parent language (PAU) and diverged from one another with the relocation of Proto-Utian speakers from south Alaska to California, while speakers of Aleut remained in the Alaska homeland.

For thousands of years, Aleut (as a distinct language) was in contact (either directly or indirectly) with other circumarctic languages, including those of a large linguistic diffusion area extending from Alaska to central Siberia, where most Altaic and Uralo-Siberian languages are spoken (cf. Fortescue 1998, Fortescue and Vajda 2022). Aleut's SVPs with West Greenlandic (Eskimo), Nenets (Uralic), and Evenki (Tungusic) are all greater in value than its SVP with Southern Sierra Miwok (table 3.2). Moreover, in California, Utian languages were influenced by structures of local languages, contributing even more to their structural differences with Aleut. Even so, as sister languages, Aleut and Southern Sierra Miwok retained enough structural similarity over time to be located in the ninety-second percentile of structural resemblance for 157 *WALS* languages (table 3.2). If Utians had not departed from south Alaska and remained in contact with, but lived separated from, speakers of Aleut, structural similarities for the two languages would likely have been substantially greater.

Chapter 4

Mitochondrial DNA Evidence

4.1 INTRODUCTION

Having examined linguistic evidence (lexical and structural) for the Aleut/Utian connection, we turn to human molecular genetics, specifically to mitochondrial DNA (mtDNA), for additional data. Mitochondrial DNA is especially important in genetic investigation because this material is inherited matrilineally, that is, through a single lineage, that of females. Using mtDNA in genetic analysis helps avoid some of the complexities of recombinant-genetics study, where hereditary material is contributed by both males and females in meiosis.

Mitochondrial DNA analysis is not the only approach to population history using molecular genetics. Contemporary geneticists now widely use both Y chromosome as well as whole genome analysis. For various reasons, however, most published works available for Native American populations are mtDNA descriptions, and we adopt the mtDNA approach here to fit this reality. Mark Stoneking, author of *An Introduction to Molecular Anthropology* (2016), in response to reading the first draft of this chapter, makes this observation:

> Although we [human geneticists] started population history work analyzing mtDNA, we have mostly moved on to analyzing genome-wide data, which is more informative for detecting population relationships . . . But I think you are justified in mentioning only the mtDNA work, because there is by far more mtDNA data than genome-wide data for North America—in fact to the best of my knowledge there is no genome-wide data available from any Utian-speaking group or from the Aleutians.

For a useful, brief introductory discussion of mtDNA in studying human prehistory, see Johnson et al. 2012:49–50.

4.2 HAPLOGROUPS

An mtDNA haplotype is a cluster of alleles in an organism inherited together from a single parent. An mtDNA haplogroup is a collection of similar haplotypes having a common female ancestor. The discussion of human genetics in this chapter and the following chapters focuses exclusively on mtDNA haplogroup evidence.

Geneticists recognize five mtDNA haplogroups for Native Americans, by which this large population of the Western Hemisphere can be distinguished genetically from all other human groups in the world. These are haplogroups *A, B, C, D,* and *X*, sometimes referred to as "founding mtDNA haplogroups" (e.g., Crawford et al. 2010:700).

Each haplogroup of a population can be distinguished by its average frequency of occurrence within the population. For example, frequencies for *A, B, C, D,* and *X* in a sample group of 115 speakers of Old Harbor Eskimo (Alutiiq) of Kodiak Island, Alaska, are 61.7 percent *(A)*, 3.5 percent *(B)*, 0.0 percent *(C)*, 34.8 percent *(D)*, and 0.0 percent *(X)* (Merriwether et al. 1995). In our presentation of Native American haplogroups (mtDNA profiles), we exclude haplogroup *X*, as is often done in such comparative studies; frequencies for *X* in most Native American groups are commonly found to be zero or very near zero.

We present mtDNA profiles using a gray scale of four degrees of contrast to indicate relative average frequencies of haplogroups (cf., Lorenz and Smith 1996:317, figure 3) (Hereafter, we substitute "relative frequency" for "relative average frequency."). For example, the relative frequencies of the four haplogroups for the population of speakers of Old Harbor Eskimo (Merriwether et al. 1995) are as presented in table 4.1.

In the table, *N* is the population sample size, and *A, B, C,* and *D* are the four respective haplogroups recognized for Native American populations. Relative frequencies of haplogroups are indicated by shading frequency numbers using four different degrees of gray. With regard to Old Harbor Eskimo, the lightest shading (darker 5 percent) indicates haplogroup *A* to be of greatest frequency among the four (*A* in the example of table 4.1); the next darker shade (darker 15 percent), indicates the second-most frequent haplogroup (*D* in the example); the next darker shade (darker 25 percent), indicates the third-most frequent haplogroup (*B* in our example); and the

Table 4.1 Two Examples of Graphics for mtDNA Profiles Extracted from Online Appendix C

Population Language	*N*	*A*	*B*	*C*	*D*
Old Harbor Eskimo (Yupik)	115	62.1	3.5	0	34.8
Bella Coola (Nuxalk)	25	60	8	8	20

Source: Generated by the authors.

darkest shade (darker 35 percent), is the least most-frequent haplogroup (*C* in our example).

The frequencies of haplogroups of a single population are sometimes numerically identical. When so, the tied haplogroups (i.e., those with identical frequencies) show the same gray shading. For example, see Bella Coola in table 4.1 (Merriwether et al., 1995) where haplogroups *B* and *C* of Bella Coola are tied for frequency and, consequently, share the same gray shade (darker 25 percent). *B* and *C* share this contrast since these two contend for being the third-most frequent haplogroup after *A* (first) and *D* (second). No single haplogroup of Bella Coola can be identified as the fourth most frequent of the four; thus, no haplogroup frequency is shaded darker 35 percent.

Haplogroup profiles of populations are compared to ascertain the degree to which populations are genetically related to one another. Two populations with similar profiles are more closely genetically affiliated with one another than two populations with less similar profiles.

Our visual presentation of the relative frequency of haplogroups makes it reasonably simple to compare profiles for similarity or difference. Consider the five comparisons (A–E) in table 4.2 based on data from online appendix C.

Table 4.2 Five Binary Comparisons of Haplogroup Profiles Based on Data from Online Appendix C

Population Language Name or Archaeological Site Name	*Score*	*N*	*Haplogroups*			
			A	*B*	*C*	*D*
Comparison A	4					
Ouzinkie Eskimo		41	73.2	0	4.9	14.6
Gambell Eskimo		50	58	0	14	26
Comparison B	3					
Western Basketmaker II		23	13	78.3	4.3	4.3
Anasazi 1		27	22	56	15	0
Comparison C	2					
Aleut (Pribilof Islands)		78	25	0	1.4	66.7
Northern Paiute		94	0	42.6	9.6	47.9
Comparison D	1					
Northern Paiute		94	0	42.6	9.6	47.9
Washo		28	0	53.6	35.7	10.7
Comparison E	0					
Old Harbor Eskimo (Yupik)		115	61.7	3.5	0	34.8
Aleut (Pribilof Islands)		78	25	0	1.4	66.7

Source: Generated by the authors.

In comparison A, all four haplogroups, *A, B, C,* and *D*, have the same relative frequencies in each of the two compared profiles. "Score" measures the number of equivalent relative frequencies observed for the compared profiles. The score for comparison A is 4. In comparison B, three of the haplogroups, *A, B*, and *C*, have the same relative frequencies, and the relative frequencies for *D* are different. The score for comparison B is 3. In comparison C, two of the compared haplogroups, *C* and *D*, are the same, and two, *A* and *B*, are different. The score for comparison C is 2. In Comparison D, only one pair of haplogroups, those of *A*, show the same relative frequencies. *B, C,* and *D* do not. The score for comparison D is 1. In comparison E, none of the four compared haplogroups have the same relative frequencies. The score for comparison E is 0.

To summarize, comparison A shows a score of 4, B shows 3, C shows 2, D shows 1, and E shows 0. The comparison score is not only indicative of profile degree of similarity, but also of associated population degree of genetic relationship, wherein the greater the score, the closer the genetic affiliation, and the smaller the score, the more distant the genetic relationship. The smallest possible difference is 0 (zero) where two profiles show no commonality with respect to relative frequency. This score does not indicate a lack of genetic relationship, only that among the possible scores, profile comparisons with zero show the most distant genetic affiliation.

4.3 MITOCHONDRIAL DNA DATABASE

Mitochondrial DNA data used in this book are presented in their entirety in online appendix C. This involves haplogroup profiles for each of the 165 Native populations distributed across North America. This information is drawn from scientific articles and chapters ranging in publication date from 1992 (Torroni et al.) to 2017 (González-Oliver et al.). Our assembly of haplogroup profiles for Native American populations is nearly, if not completely exhaustive of that available in published literature, except that our assembly does not include any population sample wherein *N* is less than sixteen. Consult online appendix C for more details.

4.4 DISTRIBUTION OF MOST-FREQUENT HAPLOGROUP *D*

Distribution of haplogroups across population languages and geographical locations is important as evidence for the investigation of human prehistory in addition to linguistics (chapters 1, 2, and 3) and archaeology (chapter 8). In this section, we focus on the distribution of haplogroup *D* when it is most frequent in a profile. In shorthand presentation, we refer to this as "most-frequent *D*."

Table 4.3 Twenty-Nine Native North American Populations, Modern and Ancient, with Most-Frequent *D*, Extracted from Online Appendix C (See Also Map of Figure 4.1)

	Region 1: Modern Arctic								
ID	*Language*	*Family*	*Location*	*N*	*A*	*B*	*C*	*D*	*Key*
1	Aleutian Islands	Aleut	Alaska	163	34.4	0	0	65.6	1
2	Living Aleuts	Aleut	Alaska	198	28	0	0	72	2
3	Aleut (Atka)	Aleut	Alaska	17	29.4	0	0	70.6	3
4	Aleut (Atka)	Aleut	Alaska	179	28.5	0	0	71.5	4
5	Aleut (Sand Point)	Aleut	Alaska	38	44.7	0	0	55.3	3
6	Aleut1	Aleut	Alaska	267	41.6	0	0	58.4	3
7	Aleut (St. George)	Aleut	Bering Sea	28	17.9	0	0	82.1	3
8	Aleut (St. Paul)	Aleut	Bering Sea	35	28.6	0	0	71.4	3
9	St. Paul Aleut	Aleut	Bering Sea	72	25	0	1.4	66.7	5
10	Aleut (Pribilof Is.)	Aleut	Bering Sea	78	25	0	1.4	66.7	4
11	Commander Island	Aleut	Bering Sea	36	0	0	0	100	1
12	Aleut2	Aleut	Bering Sea	35	0	0	0	100	3
	Region 1: Ancient Arctic								
ID	*Site Name*	*Date*	*Location*	N	*A*	*B*	*C*	*D*	*Key*
13	Aleut	NA	Alaska	17	35.3	0	0	64.7	6
14	Late pre-contact Aleut	1000–400 BP	Alaska	52	23.1	0	0	76.9	7
15	Aleutian Island	Post AD-1000	Alaska	64	23	0	0	77	2
16	Paleo-Aleut	NA	Alaska	46	42	0	0	58	2
17	Neo-Aleuts	NA	Alaska	33	33	0	0	67	2

(*Continued*)

Table 4.3 (Continued)

	Region 3: Modern Northwest Coast								
ID	*Language*	*Family*	*Location*	N	*A*	*B*	*C*	*D*	*Key*
18	Splatsin	Salish	British Columbia	19	16	5	32	32	8
	Region 4: Modern California (San Francisco Bay Area [SFBA])								
ID	*Language*	*Family*	*Location*	N	*A*	*B*	*C*	*D*	*Key*
19	Yok-Utian	Utian and Yok-Utian	Central California	17	11.8	29.4	11.8	47.1	9
20	California Penutian	Utian etc.	Central California	17	12	41	6	41	10
	Region 4: Ancient California (SFBA)								
ID	*Site Name*	*Date*	*Location*	N	*A*	*B*	*C*	*D*	*Key*
21	Yukisma SCL-38 1	2205–245 BP	SFBA	200	5.5	22	29.5	43	11
22	Sandhill Rd. SCL-287	1176–2220 BP	SFBA	23	4.3	8.7	13	74	11
23	Cook CA-SOL-270	*c. 2000 BP*	SFBA	23	4	9	44	44	11
24	Yukisma SCL-38 2	245–2200 BP	SFBA	71	9.9	22.5	21.1	46.5	12
	Region 6: Modern Great Basin								
ID	*Language*	*Family*	*Location*	N	*A*	*B*	*C*	*D*	*Key*
25	Northern Paiute	Uto-Aztecan	NW Great Basin	94	0	42.6	9.6	47.9	9
	Region 6: Ancient Great Basin								

ID	*Site Name*	*Date*	*Location*	N	*A*	*B*	*C*	*D*	*Key*
26	Pyramid Lake	860–5905 BP	West Gr. Basin	18	11.1	33.3	0	55.6	10
27	Stillwater Marsh	354–2373 BP	Carson Sink	21	4.8	38.1	0	57.1	10
	Region 9: Modern Southeast								
ID	*Language*	*Family*	*Location*	N	*A*	*B*	*C*	*D*	*Key*
28	Muskoke (Oklahoma)	Musgokean	Alabama, Georgia	71	36.6	15.5	9.9	38	5
	Region 11: Modern Middle America								
ID	*Language*	*Family*	*Location*	N	*A*	*B*	*C*	*D*	*Key*
29	Mayo	Uto-Aztecan	Mexico	55	23.6	30.9	12.7	32.7	13

Source: (indicated by key in table): 1. Volodko et al. 2008, 2. Smith et al. 2009, 3. Crawford et al. (2010), 4. Rubicz et al. 2003, 5. Merriwether et al. 1995, 6. O'Rourke et al. 2000, 7. Raff et al. 2011, 8. Hughes et al. 2016, 9. Eshleman et al. 2004, 10. Kaestle 1998, 11. Monroe 2014, 12. Johnson et al. 2012, 13. Gorostiza et. al. 2012. Data compiled by authors.

Table 4.3 presents haplogroup profiles with most-frequent *D* found among the 165 populations of online appendix C. There are twenty-nine such profiles, constituting 17.6 percent of the total number of populations (29/165), which indicates that the occurrence of most-frequent *D* is relatively rare in Native North American populations (cf. Eshleman et al. 2004:68). In contrast, the percentage for most-frequent *A* is 55.8 (92/165). The twenty-nine populations also have a relatively limited geographic distribution, with most of them found in the Arctic in south Alaska (17), in California in the San Francisco Bay Area (6), and in the Great Basin (3). Three others are in the Northwest Coast (1), Mexico (1), and the Southeast (1) (see map in figure 4.1).

Most-frequent *D* distribution is important because it coincides with the dispersal of languages of the proposed Aleut-Utian language family in southwest Alaska and in central California's San Francisco Bay Area. Moreover, to the immediate east of the latter area is the western Great Basin, where three of the 165 populations show most-frequent *D*. All three of these areas include both modern and ancient most-frequent *D* populations.

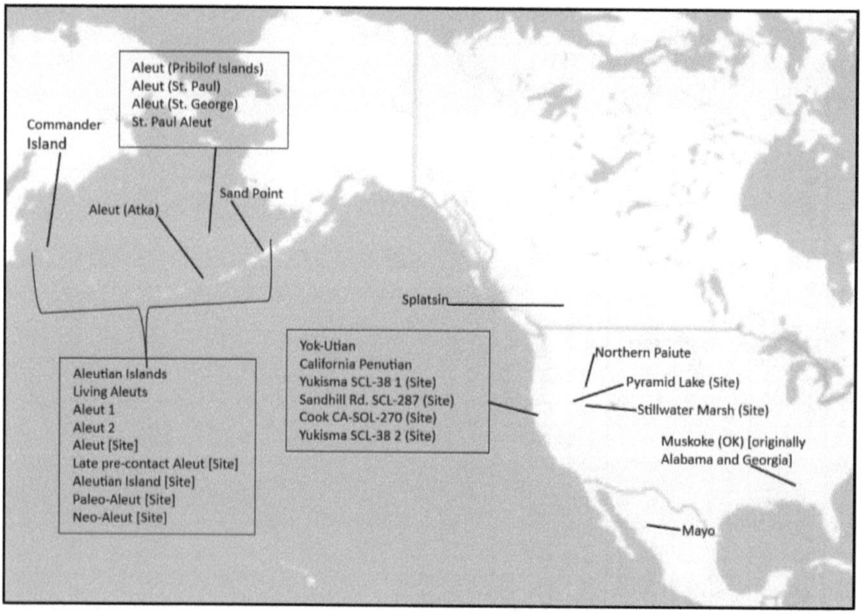

Figure 4.1 Map of North America Showing Distribution of Native Populations Having a mtDNA Profile with a Most-Frequent *D* Haplogroup (see Table 4.1). *Source*: Map outline adapted from Dryer and Haspelmath (2013). Permission granted by Martin Haspelmath.

These distributions accord with our model of Utian prehistoric migration and language contacts based on linguistic reconstruction discussed in depth in chapters 6, 9, and 10.

Comparative linguistics (chapters 1 and 2) supports the reconstruction of a language we dub PAU spoken in coastal southwest Alaska during the Middle Holocene era. Speakers of the Utian dialect of PAU migrated by watercraft from Alaska to California (specifically to the San Francisco Bay Area), leaving behind in Alaska speakers of their sister dialect, Aleut. This geographic separation of dialects resulted in the divergence of PAU into two distinct daughter languages, Proto-Utian and Early Aleut. Along the way from Alaska to California, Utian migrants came into prolonged contact with several different groups of people of coastal British Columbia and Washington State with whom loanwords were exchanged (chapter 6) (and mitochondrial DNA as well [see below discussion of Splatsin]).

In our model, Proto-Utian and Early Aleut are sibling languages, both having the same parent language, PAU. Most respective speakers of these two tongues belonged to the same genetic population (with most-frequent *D* in both groups). After migration, most speakers of Aleut who remained in Alaska and most speakers of Utian who relocated to California continued to be genetically similar by sharing most-frequent *D* (barring some developments such as total language replacement). This explains the finding reported in table 4.3 and figure 4.1 that populations of Alaska and California share most-frequent *D* haplogroups.

Aleuts and Utians also share most-frequent *D* with three populations of the North American Great Basin (online appendix C, region 6), that is, speakers of Northern Paiute (Uto-Aztecan) located in the northwest Great Basin, and members of two ancient populations, Pyramid Lake of the western Great Basin, and Stillwater Marsh at Carson Sink in western Nevada (table 4.3 and figure 4.1).

The occurrence of most-frequent *D* in populations of the western Great Basin is likely due at least in part to its geographic proximity to the San Francisco Bay Area, to which most-frequent *D* had originally been brought by a seafaring population or populations from southwest Alaska (Utians). The fact that an ancient population of the Bay Area, Yukisma SCL-38 2 (table 4.4, ID 1), and modern Northern Paiute (Great Basin) (table 4.4, ID 2) have identical relative-frequency profiles (comparison score 4) provides strong support for this conjecture.

Northern Paiute and modern Yok-Utian (table 4.4, IDs 2 and 3) of central California (inclusive of the Bay Area) also share this same profile (comparison score 4).

Profiles for the two ancient populations of the western Great Basin, Pyramid Lake, and Stillwater Marsh (Kaestle 1998) (table 4.4, IDs 4 and 5), are identical to one another (comparison score 4), and also similar to (but not

Table 4.4 Haplogroup Profiles Extracted from Online Appendix C Discussed in the Text and Listed Here in Order of Reference

ID No.	*Population Language (Family) or Archaeological Site*	*A*	*B*	*C*	*D*
1	Yukisma SCL-38 2	9.9	23	21	46.5
2	Northern Paiute (Uto-Aztecan)	0	43	9.6	47.9
3	Yok-Utian (Yokuts and Utian)	11.8	29.4	12	47.1
4	Pyramid Lake	11.1	33	0	55.6
5	Stillwater Marsh	4.8	38	0	57.1
6	Mayo (Uto-Aztecan)	23.6	31	13	32.7
7	Nahua (Uto-Aztecan)	61.3	32.3	6.5	0
8	Splatsin (Salish)	16	5	32	32
9	Muskoke (Muskogean)	36.6	15.5	9.9	38
10	St. Paul Aleut (Aleut)	25	0	*1*	67
11	Atka (Aleut)	29	0	0	71

Source: Generated by the authors.

identical with) profiles observed for Yukisma SCL-38 (ID 1), Northern Paiute (ID 2), and Yok-Utian (ID 3) (comparison score 2).

The Mayo-speaking population of Middle America (table 4.3, region 11, population 147) was also involved in the above-described mix. Mayo (Gorostiza et al. 2012) is a Uto-Aztecan language spoken in northwest Mexico (figure 4.1). Its profile (table 4.4, ID 6) is identical to that found for ancient populations of Pyramid Lake (ID 4) and Stillwater Marsh (ID 5) sites in the western Great Basin (comparison score 4).

Ancestors of modern speakers of Mayo were at one time likely occupants of parts of the western Great Basin, where this population shared haplogroup profiles with populations of Pyramid Lake and Stillwater Marsh and were possibly inhabitants of one or both of these sites. Subsequently, this population moved south to northwest Mexico, where it retained its Great Basin haplogroup profile but replaced its Great Basin language (whatever that might have been) with a Uto-Aztecan language spoken in its new habitat. Today, Mayo stands alone among its nearby Uto-Aztecan-speaking neighbors of Middle America by showing a Great Basin profile (table 4.4, ID 6) which is very different from those of neighbors such as Nahua (Uto-Aztecan) (table 4.4, ID 7) (see region 11 of online appendix C for other examples).

Another most-frequent *D* population of North America is Splatsin (table 4.4, ID 8) (Hughes et al. 2016), a Salish language group of the Northwest Coast, British Columbia (figure 4.1). This population's profile shows a gray shading (darker 5 percent) indicating most-frequent *D*, but its *C* haplogroup also shows the same shading indicating most-frequent C, with both *C* and *D* competing to be the haplogroup with the greatest frequency (32 percent). Nevertheless, the Splatsin population is judged as having most-frequent *D*, a feature possibly acquired through admixture with Utian mariners during a prolonged pause in their maritime journey (or journeys) from coastal southwest Alaska to the San Francisco Bay Area (figure 4.1).

Most-frequent *D* is also found in a population of the Southeast (figure 4.1) speaking Muskoke (Muskogean) (Merriwether et al. 1995). Its profile is given in table 4.4, ID 9. No profile is exactly the same as this among the other twenty-eight in table 4.3. The closest are seventeen populations, modern and ancient, of the Arctic region (1), all with the same near-identical profile to that of Muskoke. For example, see table 4.4, IDs 10 and 11. All modern populations with similar profiles (i.e., most-frequent *D* and second-most frequent *A*), save Muskoke, speak Aleut.

While outside the scope of this study, just for the record, twenty-eight Indigenous Populations of South America, modern and ancient, also show haplogroup profiles with most-frequent *D*.

Chapter 5

Other Relationship Proposals for Aleut and Utian

5.1 INTRODUCTION

The genealogical relationship for Aleut and Utian proposed here has implications for other proposals connecting these languages to other language groups. Costanoan and Miwok, spoken in central California, were proposed by Dixon and Kroeber (1919) to be joined with Yokuts, Maidu, and Wintu (all also of California) in "California Penutian" (Sapir 1929). (The close relationship of Miwok and Costanoan languages in Utian was not established until later.) California Penutian has held up reasonably well over the past century, despite various challenges to its validity (e.g., Shipley 1983–1984). The relationship of this language group to other languages outside of California has been proposed, forming subgroups of Penutian such as "Plateau Penutian" (interior Oregon and Washington State), "Oregon Penutian" (coastal Oregon and Washington State), and "Mexican Penutian" south of the border, considerably extending geographically possible relationships for Utian.

The most widely acknowledged proposal for Aleut is kinship with Eskimo (Inuit and Yupik languages of northern North America, Alaska, Canada, Greenland, and extreme eastern Siberia). An Eskimo-Aleut connection has long been recognized by scholars (Rask 1820, see Thalbitzer 1921). More recently, Eskimo-Aleut has been proposed by Fortescue (1998) as a New World component of "Uralo-Siberian." This group includes Uralic (Europe and Asia) and Yukaghir (Siberia) and, originally, Chukotko-Kamchatkan, but Fortescue has now removed this east Siberian group from Uralo-Siberian (2011:136; Fortescue and Vajda 2022).

In tables 5.1 and 5.2, we evaluate lexical evidence for these various proposals using the quantitative approach described in chapter 1. Table 5.1 treats

Utian and California Penutian and 5.2 treats Aleut and Uralo-Siberian. In the tables, Utian and Aleut languages are lexically compared with languages for which a relationship with one or the other of these has been proposed. We extract evidence for these comparisons from the 232 comparisons presented in online appendix A. Table 5.1 presents results for nineteen binary comparisons, in each of which one language is California Penutian, including seven Utian comparisons each matched with a California Penutian language. Table 5.2 gives fifteen binary comparisons, for each of which one language is Aleut, including seven Aleut comparisons matched with a Uralo-Siberian language. See chapter 1 (1.2) for full details of the quantitative method used to produce results reported in tables 5.1 and 5.2, including abbreviations and conventions used in the tables (table 1.1).

5.2 QUANTITATIVE EVIDENCE

The overall strength of support for the historical connection of languages proposed as related to Utian and Aleut languages is described in the final (far right) columns of tables 5.1 and 5.2 and in sections 5.3 and 5.4, respectively. These involve prosaic descriptions of evidential support based on quantitative evaluation (1.2.4).

5.3 CALIFORNIA PENUTIAN

5.3.1 Quantitative Screening and California Penutian

The first comparison in table 5.1 (beginning with the top data row) is Proto-Costanoan/Proto-Miwok with *Strong for Kinship*. Both languages are Utian, and their relationship defines the Utian group. This is the only comparison of the table with *Strong* support for a connection. The next five comparisons show *Moderate for Kinship,* indicating these languages could be conclusively shown to be genealogically affiliated after further study. This is precisely the result for Aleut/Proto-Miwok (repeated in table 5.2), to which the traditional comparative method is applied in chapter 2. Two *Moderate for Borrowing* comparisons follow. With the exception of three *Moderate for Connection* comparisons, the final eleven are either *Weak for Borrowing* (one) or *No Support for Connection* (seven), none of which are worthy of additional comparative investigation.

Figure 5.1 illustrates graphically the relationships reported in table 5.1. These are the ten comparisons of table 5.1 whose results include *Moderate for Kinship*, *Moderate for Borrowing*, or *Moderate for Connection*. The remaining seven are either *Weak for Borrowing* (1) or *No Support for Connection* (6)

Table 5.1 Evaluation of Evidence for Nineteen Binary Comparisons (From the Sample of 232, Online Appendix A), Each of Which Entails at Least One California Penutian Language

ID No.	*Language Comparison*	*Possible Group*	*BV%*	*NBV*	*ASJP*	*Strength of Support*
19	Proto-Costanoan/Proto-Miwok	California Penutian (Utian)	*47.1*	*16*	*7.83*	Strong for Kinship
56	Aleut/Proto-Miwok	Aleut-Utian	*35.7*	*10*	−0.10	Moderate for Kinship
60	Aleut/Proto-Yokuts	Aleut-Yokuts	*50*	*8*	−2.69	Moderate for Kinship
69	Proto-Maidun/Proto-Yokuts	California Penutian	*33.3*	1	*4.67*	Moderate for Kinship
72	Wintu (Wintun)/Lake Miwok (Utian)	California Penutian	*33.3*	4	*4.2*	Moderate for Kinship
79	Takelma/Yokuts	Penutian	*66.7*	2	2.43	Weak for Kinship
121	Maidu (Maidun)/Lake Miwok (Utian)	California Penutian	14.3	2	*5.22*	Moderate for Borrowing
127	Yuki/Yawelmani Yokuts	Yukian-California Penutian	0	0	*3.68*	Moderate for Connection
135	Proto-Maidun/Proto-Wintun	California Penutian	0	0	*6.26*	Moderate for Connection
143	Proto-Yokuts/ Proto-Miwok	California Penutian	0	0	*4.24*	Moderate for Connection
144	Alsea/Wintu (Wintun)	Alsea/Wintun (< Penutian)	0	0	*4.1*	Moderate for Connection
145	Wintu (Wintun)/ Southern Sierra Miwok	California Penutian	12.5	4	3.13	Weak for Borrowing
147	Zuni/Yawelmani Yokuts	Zuni-California Penutian	23.1	3	2.75	Weak for Borrowing
166	Maidu (Maidun)/ Yawelmani Yokuts	California Penutian	0	0	2.62	No Support for Connection
180	Wintu (Wintun)/ Yawelmani Yokuts	California Penutian	0	0	2.03	No Support for Connection
183	Coos/Yokuts	Penutian	0	0	1.83	No Support for Connection
198	Aleut/Proto-Wintu	Aleut-Wintu	0	0	0.76	No Support for Connection
223	Takelma/Maidu	Penutian	0	0	−0.99	No Support for Connection
232	Proto-Eskimo/Proto-Miwok	Eskimo-Utian	0	0	−7.5	No Support for Connection

Source: Generated by the authors.
Note: Italicized values are all equal to or greater than average (mean *BV%* = 28.42, mean *NBV* = 6.6, mean *ASJP* = 3.36).

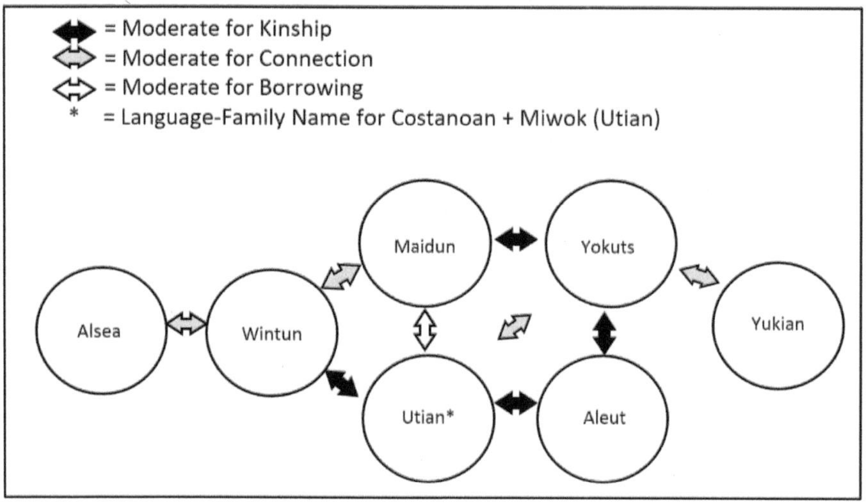

Figure 5.1 Relationship Map for California Penutian Based on Results Reported in Table 5.1. *Source*: Generated by the authors.

and are not included on the relationship map. The single *Strong for Kinship* is also not included on the chart.

All four language groups of California Penutian—Utian (Costanoan and Miwok), Maidun, Wintun, and Yokuts—are chained together through the relationships *Moderate for Kinship*, *Moderate for Connection*, and *Moderate for Borrowing* (respectively, double-headed black arrows, double-headed gray arrows, and a double-headed white arrow). Links in this chain are transitive (except *Moderate for Borrowing*) such that, for example, Wintun could be genealogically related to Yokuts through its affiliation with intervening Maidun, and vice versa.

This relational complex also extends to three non-California Penutian languages: Aleut, Alsea, and Yukian (cf., DeLancey and Golla 1997). Alsea could belong to California Penutian if further comparative study confirms its connection with Wintun (double-headed gray arrow), and Yukian could belong to California Penutian if further comparative study confirms its link with Yokuts. Only one of these three, Aleut, is linked by *Moderate for Kinship* to California Penutian languages. Aleut is connected to both Utian and Yokuts (double-headed black arrows).

Wintun and Utian (i.e., comparison 72, table 1.5, Wintu/Lake Miwok) are linked through *Moderate for Kinship* (double-headed black arrow). This connection should be checked through further research, but our prediction is that additional evidence will not bear witness to a kinship affiliation, since evidence for Wintun/Southern Sierra Miwok (145 [table 5.1]) is *Weak for Borrowing* rather than *Moderate for Kinship*.

Of special note on the map is the direct kinship affiliation of Yokuts and Aleut. If this relationship were to be maintained after further study, then speakers of Proto-Yokuts might have been involved in a migration similar to that of speakers of Proto-Utian from Alaska to California. We have pursued further study of this connection using the traditional comparative method and have not been able to confirm a relationship between Yokuts and Aleut.

The relational map in figure 5.1 should not be mistaken for a language-family tree model for California Penutian. Unlike a family tree, the figure does not chart hierarchical levels of language inclusion. Rather, it merely indicates what languages are possibly related to one another and the quality of support provided for these relationships by evidence. Relational maps for languages might be used as aids in constructing trees for language families, but only if the relationships charted for all languages on a map are understood to be conclusive in one way or another (i.e., through kinship, borrowing, or not at all). Therefore, a relational map such as that in figure 5.1, based solely on the results of initial quantitative assessments of language affiliations, should be considered decisive only after each relationship has received additional comparative study using a traditional non-quantitative approach, such as the comparative method of historical linguistics. Nevertheless, even if not conclusive, first-stage relational maps provide useful provisional insights for plotting the way forward in research.

5.3.2 *ASJP World Tree* and California Penutian

The ASJP World Language Tree of Lexical Similarity graphically illustrates relative degrees of lexical resemblance among 7,996 of the world's languages and dialects through the use of a family-tree-like model and essentially represents another perspective on ASJP classificatory results. In the tree, doculects most lexically similar to one another (determined through the ASJP algorithm, 1.2.3) are branched more closely together and those less similar to one another are branched further apart, with tree branches ultimately joining all 7,996 doculects together on a single trunk (Müller et al. 2021, The ASJP Database - (clld.org)).

World Tree results for California Penutian show Maidun and Utian languages branched closely together, while no other California Penutian languages are joined so closely to either of these (figure 5.2). Figure 5.1 also indicates a close lexical similarity between Maidun and Utian, this based on borrowing rather than kinship.

The *World Tree* fragment of figure 5.2 also shows close branching between Yukian languages (Yuki and Wappo) and California Penutian languages (Maidun and Utian), a resemblance also observed on the relationship map for California Penutian of figure 5.1 with the evaluation of *Moderate for Connection* for Yukian and Yokuts (California Penutian).

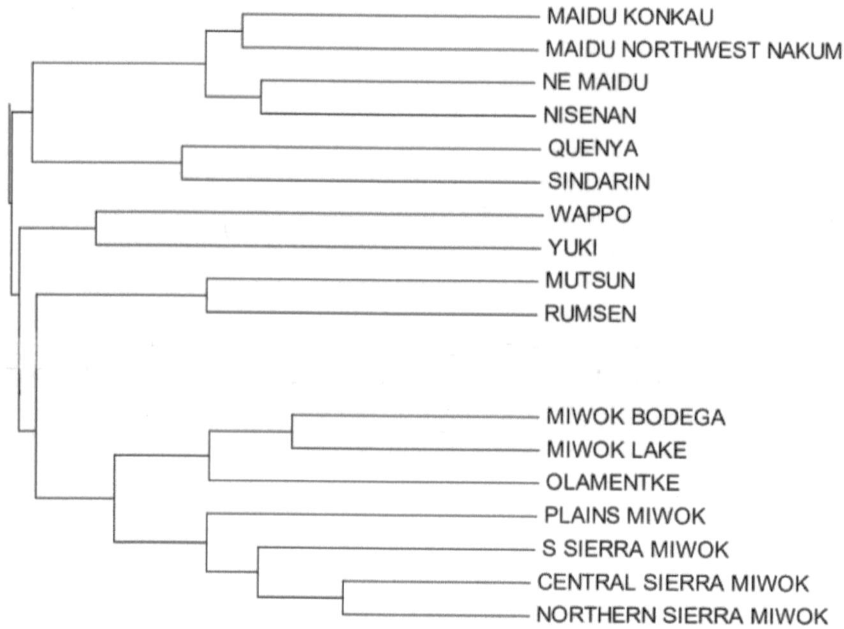

Figure 5.2 Maidun and Utian Languages on a Fragment of the ASJP World Tree. *Note*: Maidun = Maidu Konkau to Nisenan; Utian = Mutsun to Northern Sierra Miwok; all other languages are non-California Penutian. *Source*: Müller et al. (2021). Permission granted by Cecil H. Brown.

5.4 URALO-SIBERIAN

5.4.1 Quantitative Screening and Uralo-Siberian

Table 5.2 presents evaluative results for fifteen binary comparisons, each of which entails a Uralo-Siberian language as recognized by Fortescue. In Fortescue and Vajda (2022), Fortescue distinguishes four Uralo-Siberian languages: Proto-Uralic, Proto-Yukaghir, Proto-Eskimo, and Aleut. Also included in the table are Proto-Samoyedic (offspring of Proto-Uralic) and Proto-Inuit (offspring of Proto-Eskimo). Non-Uralo-Siberian languages found in comparisons are Proto-Miwok (Utian), Proto-Wintu, Proto-Yokuts (California Penutian), and Kwak'wala (Wakashan).

Figure 5.3 graphically illustrates some of the relationships reported in table 5.2. These are the seven comparisons in the table, the results of which include *Moderate for Kinship* (three comparisons), *Moderate for Borrowing* (three), and *Moderate for Connection* (one). The remaining eight comparisons show either *Weak for Kinship* (two) or *No Support for Connection* (six) and are not included on the relationship map.

Table 5.2 Quantitative Evaluation of Evidence for Fifteen Binary Comparisons (From the Sample of 232, Online Appendix A), Each of Which Entails at Least One Uralo-Siberian Language

ID No.	*Language Comparison*	*Possible Group*	*BV%*	*NBV*	*ASJP*	*Strength of Support*
56	Aleut/Proto-Miwok	Aleut-Utian	*35.7*	*10*	−0.10	Moderate for Kinship
60	Aleut/Proto-Yokuts	Aleut-Yokuts	*50*	*8*	−2.69	Moderate for Kinship
77	Proto-Samoyedic (Uralic)/Proto-Yukaghir	Uralo-Siberian	*50*	3	*3.52*	Moderate for Kinship
102	Proto-Samoyedic (Uralic)/Proto-Eskimo	Uralo-Siberian	*28.6*	4	−0.8	Weak for Kinship
103	Proto-Eskimo/Proto-Yukaghir	Uralo-Siberian	*50*	2	−1.16	Weak for Kinship
115	Proto-Eskimo/Aleut	Uralo-Siberian	*16*	*16*	0.9	Moderate for Borrowing
116	Aleut/Proto-Inuit (Eskimo)	Uralo-Siberian	*18.8*	*13*	0.84	Moderate for Borrowing
119	Proto-Uralic/Proto-Dravidian	Nostratic	26	26	−0.91	Moderate for Borrowing
128	Proto-Uralic/Proto-Yukaghir	Uralo-Siberian	0	0	*3.65*	Moderate for Connection
172	Proto-Eskimo/Kwak'wala	Eskimo-Aleut-Wakashan	0	0	2.8	No Support for Connection
186	Proto-Yukaghir/Aleut	Uralo-Siberian	0	0	1.43	No Support for Connection
198	Aleut/Proto-Wintu	Aleut-Wintu	0	0	0.76	No Support for Connection
221	Proto-Uralic/Proto-Eskimo	Uralo-Siberian	0	0	−0.83	No Support for Connection
222	Proto-Uralic/Aleut	Uralo-Siberian	0	0	−0.95	No Support for Connection
232	Proto-Eskimo/Proto-Miwok	Eskimo-Utian	0	0	−7.5	No Support for Connection

Source: Generated by the authors.
Note: Italicized values are all equal to or greater than average (mean *BV%* = 28.42, mean *NBV* = 6.6, mean *ASJP* = 3.36).

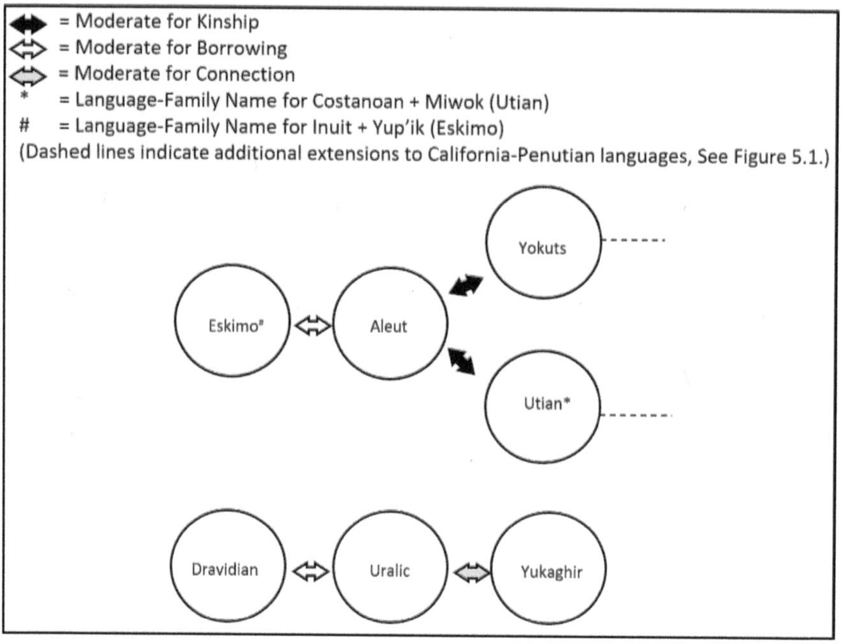

Figure 5.3 Relationship Map for Uralo-Siberian Based on Results Reported in Table 5.2.
Source: Generated by the authors.

According to quantitative findings (table 5.2) and as depicted on the relationship map of figure 5.3, we observe support for parts of Uralo-Siberian. The strongest support is for Uralic and Yukaghir, with *Moderate for Connection*. Unexpectedly, Uralic and Dravidian are linked through *Moderate for Borrowing*. Fortescue's Uralo-Siberian does not extend to Dravidian but does include Eskimo and Aleut, also linked by *Moderate for Borrowing*. If the Uralic-Yukaghir connection entails borrowing rather than kinship (a possibility provided by *Moderate for Connection*), all possible links found in table 5.2 and figure 5.3 for Uralo-Siberian involve borrowing and none involve genealogical affiliation.

5.4.2 *ASJP World Tree* and Uralo-Siberian

The ASJP World Tree result for Uralo-Siberian (figure 5.4) shows Yukaghir languages and Samoyedic languages of Uralic branched closely together, while no other languages are joined so closely to either of these except non-Samoyedic Uralic languages not included in the tree fragment of figure 5.4. Figure 5.3 also indicates a close lexical similarity between Yukaghir and Uralic involving a *Moderate for Connection* linkage. (For additional discussion of Uralo-Siberian languages, see section 5.6.)

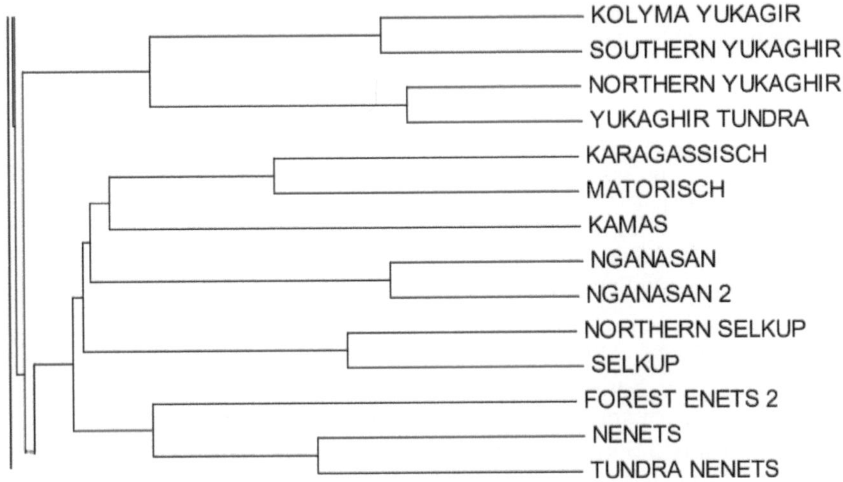

Figure 5.4 Yukaghir and Uralic Languages on a Fragment of the ASJP World Tree. *Note*: Yukaghir = Kolyma Yukagir to Yukaghir Tundra; Uralic [Samoyedic] = Karagassisch to Tundra Nenets. *Source*: Müller et al. (2021). Permission granted by Cecil H. Brown.

5.5 ESKIMO-ALEUT

Lack of genealogical relationship for Eskimo and Aleut may be the most contentious proposal of this book. Other than for borrowing, no quantitative support is found for an Eskimo-Aleut connection, or for either Eskimo or Aleut linked with a language or language group of Uralo-Siberian, and only Aleut (but not Eskimo) is supported as linked to California Penutian.

Berge (2018) undertakes a systematic evaluation of proposed cognates between Aleut and Eskimo languages extracted from the works of Bergsland (1986, 1994) and Fortescue, Jacobson & Kaplan (2010). (The latter is also the source of proposed cognates for Aleut/Proto-Eskimo to which we apply quantitative screening in chapter 1 [table 1.1, comparative set 115].) Many of these cognates are problematic, with irregular sound and/or semantic correspondences suggesting the possibility of extensive borrowing between the two languages (Berge 2018:231). Berge's task is not to challenge the Eskimo-Aleut hypothesis in any direct manner; rather, she focuses on problems involving the proposed cognates (p. 239).

Berge sorts the corpus of proposed cognates into four categories: group 1: "exact cognates," where corresponding sounds and meanings are exactly the same; group 2: "regular phonological correspondences," typical of true cognates for which phonological differences are expected; group 3: "irregular phonological correspondences," not typical of true cognates; group 4:

Table 5.3 Statistical Results of Berge's (2018:266) Sorting of 550 Possible "Cognates" for Aleut/Proto-Eskimo into Four Categories

Group	*Type of "Cognate"*	*Total*	*Questionable*	*% Questionable*
1	Exact cognates	40	20	50
2	Regular phonological correspondences	280	90–100	32–36
3	Irregular phonological correspondences	130	100	77
4	Odd semantic correspondences	100	100	100
Grand results		550	310–320	56–58

Source: Adapted from Berge (2018: 266).

"odd semantic correspondences," where compared meanings are troublingly dissimilar. Table 5.3 presents the detailed results of Berge's compilation.

Berge's (2018:266) general evaluation of these results is as follows:

> While percentages for groups 2–3 are unsurprising, the percentage of questionable "exact cognates" is noteworthy, as many of these cognates have been considered among the most illustrative of an old EA [Eskimo-Aleut] relationship. Cognates with no phonological or semantic differentiation seem improbable over the course of 3000–4000 years [i.e., the usually accepted period during which Aleut and Proto-Eskimo diverged].

It is tempting to regard the 310–320 questionable cognates as loans between Aleut and Eskimo and, by focusing on basic vocabulary, the results of our quantitative analysis would support this position. Berge is careful not to ignore other possibilities, including false cognates and language-internal processes, urging more thorough investigation of such factors (2018:265). She (p. 266) writes in a concluding observation, "This study shows that many proposed EA cognates are surprisingly similar given their non-basic nature and the assumed period of diversification of UT [Aleut] and Eskimo and that many proposed cognates could plausibly be interpreted as borrowings between the branches of the language family." She does not question approximately 35 percent of "cognates" in her compilation procedure. Nevertheless, despite expected sound correspondences and semantic equations, the latter could also result from borrowing, given the strong tendency for borrowing between Aleut and Eskimo documented by Berge and also found here through quantitative evaluation.

With respect to genetics, Rubicz et al. (2003:809) write, "Haplogroup D occur[s] at a very high frequency in Aleuts, and this, along with their unique HVS-I sequences, distinguish[es] them from Eskimos, Athapaskan Indians, and other northern Amerindian populations." Our mitochondrial

Table 5.4 Haplogroup Profiles of Discussed Eskimo and Aleut Populations

ID	*Population (All Modern)*	*A*	*B*	*C*	*D*
1	Eskimo (average of 12)	77.7	0.3	1.8	16.9
2	Aleut (average of 15)	31.4	0.0	0.2	67.5
3	Aleut (King Cove)	51.5	0	0	48.5
4	Aleut (Nelson Lagoon)	56.3	0	0	43.8
5	Aleut (Unalaska)	60.7	0	0	39.3

Source: Generated by the authors.

DNA evidence seems to provide somewhat similar low support for an Aleut/Eskimo genetic connection (comparison score 2, table 5.4, IDs 1 and 2 compared).

Among our collection of 165 Native North American Populations with haplogroup profiles (online appendix C), twelve are modern Eskimo populations, and fifteen are modern Aleut populations (region 1). The average haplogroup frequencies for these respective populations (with relative frequency indicated by degrees of gray shading) are given in table 5.4, IDs 1 and 2.

For IDs 1 and 2 (table 5.4) average *B* and *C* relative frequencies are the same, with *B* showing darker 35 percent shading for both Eskimo and Aleut, and *C* showing darker 25 percent shading for both. In contrast, *A* and *D* frequencies are reversed for Eskimo and Aleut populations, with darker 5 percent shading for *A* of Eskimo and darker 5 percent shading for *D* of Aleut, and darker 15 percent shading for *D* of Eskimo and darker 15 percent for *A* of Aleut.

The relative-frequency comparison score for Eskimo (ID 1) and Aleut (ID 2) profiles is 2 (see 4.2). Thus, the mtDNA profiles for this match are only mid-range similar. This might be considered an expected outcome for two populations like Eskimo and Aleut that originally had respective profiles in which no relative frequencies of compared haplogroups were the same, but where the two populations have been in geographic proximity for great periods of time and original mtDNA profiles have grown in some similarity through admixture.

Perhaps surprisingly, three of the fifteen modern Aleut-speaking populations show haplogroup profiles characteristic of Eskimo populations rather than of Aleut groups (table 5.4, IDs 3, 4, and 5).

We propose that this disjunct resulted from a language shift whereby an Eskimo-speaking population totally replaces its language with Aleut. In this situation, the originally Eskimo-speaking group retains its Eskimo haplogroup profile while substituting the Eskimo language with the Aleut language. For this to occur, speakers of Eskimo and Aleut must have been in close geographic proximity, probably in the general area where Aleut of the

East Aleutians (Unalaska) and on the Alaska Peninsula (King Cove and Nelson Lagoon) are spoken. This location is not too distant from the area where we hypothesize Proto-Eskimo and Proto-Aleut-Utian or Early Aleut were in close contact and exchanged large numbers of loanwords over great expanses of time (chapter 9, 9.4).

Language shift had the effect of producing a situation in which two different groups of people living in the same area, one speaking Aleut and the other speaking Eskimo, show exactly the same haplogroup profiles. Similar conditions in other parts of the Aleut-Eskimo contact area may have been responsible for the development of similar haplogroups (some identical in the past and some only partially so today) for speakers of Aleut and Eskimo languages.

5.6 FURTHER DISCUSSION OF URALO-SIBERIAN

While Fortescue now seems to acknowledge (pers. com.) a lack of strong lexical evidence for his proposal, in his recent book (Fortescue and Vajda 2022), Fortescue nonetheless assembles lexical evidence for the proposal in the form of presumed cognates supporting sound correspondences (2022) across all Uralo-Siberian languages (Uralic, Yukaghir, Eskimo, and Aleut). These are the same "cognates" (Fortescue, pers. com.) to which we apply our quantitative approach in this work, with results failing to accord with possible connections for any of the Uralo-Siberian languages except Uralic and Yukaghir (see table 5.2 and figure 5.3).

We tentatively accept a version of Fortescue's proposal in which Uralo-Siberian grammatical similarities reflect extensive contact and borrowing of great chronological depth among speakers of these languages. As such, Uralo-Siberian is a "linguistic area" or *Sprachbund* (Fortescue uses the label "linguistic mesh", cf. Piispanen 2023). A *Sprachbund* is a group of languages with many similarities in a circumscribed geographic area, most of which are not genealogically related (or if so, only very remotely).

In Fortescue's view, grammatical evidence supports Uralo-Siberian genealogical unity, even if lexical evidence (his "cognates") does not. Consequently, much of his (Fortescue and Vajda 2022) discussion of Uralo-Siberian similarities focuses on another line of linguistic evidence, grammatical morphemes (chapter 10, 10.4.2). Fortescue's work follows the waning conjecture that grammatical similarity is more probative of genealogical relationship than is lexical similarity (Campbell and Poser 2008).

The similarity of grammatical morphemes of the type cited by Fortescue (Fortescue and Vajda 2022) for Uralo-Siberian is either due to language kinship, to borrowing, or to chance (and very rarely also to language universals, not considered here). In this book, Fortescue comes close to ignoring the influence of chance, relegating mention of the topic to only these five brief statements:

> I have fleshed out the Eskimo-Uralic hypothesis lexically with many more Proto-Uralo-Siberian cognate sets since Bergsland's time and tidied up earlier morphological comparisons, rendering it highly unlikely that chance or mere borrowing lies behind it. (p.32)

> Time will tell if some of my correlations will have to be revised or rejected as due to chance. (pp. 43-44)

> Although the restricted number of cognates I have assembled may not constitute final "proof" of a genealogical relation between these languages according to the strict demands of historical linguists, they are at least sufficient to rule out chance. (p. 141)

> However, some (if not all) of these could be loanwords and some the result of chance similarities. (p. 144)

> these potential cognates—or look-alikes—are all due to chance (and there are enough of them for this to be unlikely). (p. 146).

All five comments address chance and its relation to similarity. The influence of chance on the development of morphological resemblance is mentioned only *once* by Fortescue in the first quotation cited above (Fortescue and Vajda 2022), which is curious since grammatical morphemes are considerably more likely than regular referential words to be similar by chance. Morphological words are typically less complex (shorter) than other words, thus promoting the development of coincidental similarity across languages (10.4.2). Campbell (1998:322), for example, writes, "Monosyllabic words (composed of a single consonant and vowel) may be true cognates, but they are also so short that their similarity to forms in other languages could also easily be due to chance." Like the little pig that built its house of straw, constructing a language family based on the similarity of grammatical morphemes is a precarious undertaking, especially without other resources such as lexical cognates supported by solid sound correspondences.

Chapter 6

Along the Way to San Francisco Bay

Loanwords

6.1 INTRODUCTION

The over-water distance from Anchorage, Alaska (just north of the proposed Proto-Aleut-Utian homeland, chapters 8 and 9) to San Francisco Bay, following the Pacific coast, is about 5,000 km. How long it took Utian speakers of the Middle Holocene (9.2) to make this trip by watercraft is unknown. The first intentional row across the Atlantic in 1896 took two men 55 days to travel from Manhattan to the Isles of Scilly (6,020 km), averaging 110 km per day (*oceanrowing.com*). We do not know the size of their seacraft or if Utians were rowing, paddling, or sailing; how many stops for food and water or for other purposes were made; or even how many boats were involved, among many unknowns regarding this journey. It is unlikely, however, that they similarly averaged 110 km per day. A more realistic assessment is a much slower pace, perhaps at most somewhere on the order of one-tenth that of the first Atlantic row, at an average of 11 km per day. If so, the 5,000 km journey along the coast would have taken at least 455 days.

The trip would have taken even longer if unanticipated stops were made. Such layovers could have been short, casual visits with coastal groups, or more prolonged visits with people interested in serious social intercourse (e.g., marriage, trade, or conflict). If a long interaction, archaeologists might uncover some material remains thereof, but most of, if not all such evidence would have perished over time (chapter 8). In this chapter we investigate a different kind of evidence for interaction: loanwords.

The greater the duration of contact, the more likely groups speaking different languages might exchange words. As speakers of Proto-Utian navigated along the coast, they likely encountered any number of groups speaking different languages. If a meeting lasted no more than a day or two, lexical borrowing would have been unlikely; however, if the time together were considerably more extended, by weeks or perhaps months, exchange of some vocabulary might have taken place, especially if, for example, intimate circumstances such as inter-group marriages took place and some degree of bilingualism developed. If some Utians chose to abandon their journey and stay behind, Utian words might have been even more likely to enter the vocabulary of the encountered group than vice versa. In brief, different sociolinguistic situations could have influenced the nature and extent of borrowing (Brown 1999, Thomason and Kaufman 1988).

This chapter presents loanword evidence for Utian contact with speakers of other languages as they traveled along the coast from southwest Alaska to the San Francisco Bay Area. How many different language groups could have been encountered is unknown; however, we have a good understanding of what historic languages have been and are now spoken along the way. Scanning Ives Goddard's (1999) map locating all documented North American Native languages, we find fifty-one languages, plus or minus one or two (not counting dialects), distributed along the coastal route. These are all coastal languages, with geographic ranges at least partially abutting Pacific Ocean waters or very near to the sea. Many of these languages are likely descendants of those spoken millennia ago when Utian speakers plied their way to California. In this study, we observe thirty-three Proto-Utian words possibly borrowed into vocabularies of ancient coastal languages and retained in lexicons of their modern descendants.

6.2 IDENTIFICATION OF LOANWORDS

Thirty-three historic (modern) coastal languages are consulted in this study, excluding languages from the total of fifty-one for which no (or very little) information is available. The ideal way to undertake this investigation would be to compare every Proto-Utian word reconstructed by Callaghan (2014) in her book, *Proto Utian Grammar and Dictionary*, and every Proto-Utian word reconstructed herein (chapter 2) with every meaning-equivalent word found in every lexical source (mostly dictionaries) for each of the thirty-three languages. One advantage of this approach is recognition of loanwords bidirectionally, that is, words from coastal languages borrowed into Utian and vice versa. Its major drawback is the enormous scope of this undertaking, rendering it impractical for serious consideration for the present study.

We adopt a slimmed-down, more doable strategy for finding loanwords. Instead of using all available Proto-Utian words in comparisons, only Proto-Utian reflexes of the sixty-six PAU words reconstructed in chapter 2 (2.4.6) are used in loanword searches. While this approach greatly expedites the investigation, it only facilitates a unidirectional search for loanwords, that is, loans from Proto-Utian into coastal languages are recognized, but not vice versa. This streamlined approach, then, presents a one-sided perspective on lending/borrowing activity involving Proto-Utian/other language encounters.

Loanwords are reasonably easy to identify if two compared words are both semantically and phonologically similar and are from genealogically unrelated languages. In searches, for the most part we consider as potential loans only words that are identical or very close in meaning. Phonological similarity is more complicated, however. When compared words show three or more phonological segments that are exactly the same, one of these words is likely a "copy" of the other, meaning that it is a loanword based on the other word. We dub these two words a "loan pair." When a language borrows a word from another language, the loanword typically can become phonologically altered over time as it adapts to the phonological system of the adopting language. This process makes it difficult to recognize two words as being a loan pair. To overcome this difficulty, we develop rules to follow in determining word similarity due to borrowing:

Rule 1. This rule is mentioned above: only words with the same or a closely related meaning (e.g., "sky" and "cloud") are considered potential loans.

Rule 2. At least two consonants of two compared words are phonologically similar. The compared consonants must be found in the same order in both words, for example, C_1VC_2 or C_1C_2 of one word compared to C_1VC_2 or C_1C_2 of another word is accepted (where C represents a consonant and V a vowel), but C_1VC_2 or C_1C_2 compared to C_2VC_1 or C_2C_1 is not (despite occasional transposition of sounds in words known as metathesis: e.g., *aks* for *ask* in English).

Rule 3. Judgment of phonological similarity is based on principles of articulatory phonetics, where the articulatory difference for compared segments typically entails no more than one distinguishing feature. "Articulatory" refers to where and how sounds of languages are produced by vocal organs (mouth, tongue, nose, and throat). For example, *k* and *g* are articulatorily alike (both are velar stops) but differ since the former is unvoiced and the latter voiced; *ts* and *tʃ* are alike (both unvoiced affricates) but differ since the former is alveolar and the latter is palatal/alveolar.

Rule 4. Comparisons limited to two segments are not considered indicative of a loan pair unless the segments form a consonant cluster in least one of the compared words, for example, C_1C_2. This serves to reject all two-segment comparisons consisting only of a consonant and a vowel, such as CV or VC, since similarities found for comparisons of short words are apt to have developed coincidentally rather than as products of non-chance factors such as language contact (chapter 10, 10.4.2).

In recognizing loan pairs, we follow the four rules listed above closely, but may relax one of them occasionally when circumstances clearly call for doing so; for example, rule one is relaxed when two words are phonologically very similar, but their semantic comparison is not especially close, for example, "boat hatch" and "row."

For all loan pairs we recognize, similarities are found in the word onset position. This is because the languages compared are all strongly suffixing (see chapter 3), wherein the head (root) element of the word bearing the word's semantic load is stem initial, and all modifying elements (inflections, derivatives, and such) are suffixed to the head (either directly or indirectly). Over long periods of time, head elements of words tend to be more resistant to replacement than are their affixes.

6.3 POTENTIAL LOANS INTO COASTAL LANGUAGES

Table 6.1 reports the results of searching coastal languages for words borrowed from Proto-Utian. Borrowed words are identified by resemblance to Proto-Utian (PU) words from among the sixty-six reconstructed in chapter 2, 2.4.6. Column one (from left to right) presents in geographic order names of the thirty-three languages in which potential loans are found as they are distributed along the North Pacific coast from the Alaska Peninsula to southern California (see map, figure 6.1). Four languages of the thirty-three (Esselen, Salinan, Ineseño, and Luiseño) are found south of the presumed terminal point of the Utian journey (the San Francisco Bay Area). These languages are included to ascertain whether the influence of Proto-Utian on coastal languages might have been more extensive than we have presumed (perhaps). Column two identifies the language-group classification for each of the thirty-three languages. Column three indicates the number of potential Proto-Utian loans found in each of the thirty-three languages. (These are described as "potential" since they may not be real loans, but rather chance look-alikes, see discussion below.) Column four identifies sources of lexical items for each of the thirty-three languages.

The average number of potential loans from PU for the thirty-three languages in table 6.1 is 2.2. Numbers range from nine with zero loans to one with seven (Quileute). The potential loans of table 6.1 are given in their original orthography in online appendix D.1, where presumed loans based on the same PU word are presented together in thirty-three respective charts.

Figure 6.1 provides a map locating the thirty-three historic coastal languages listed in table 6.1, giving for each its number of potential loans from Proto-Utian. Also included are locations of Aleut in Alaska and of Proto-Utian in California. Four of the thirty-three languages have five or more

Table 6.1 Number of Potential Loanwords Borrowed from Proto-Utian by Each of Thirty-Three Languages Distributed Along the Pacific Coast from the Alaska Peninsula (Alutiiq) to Southern California (Luiseño)

Language (Roughly Along the Pacific Coast from North to South)	*Language Classification*	*Total Potential PU Loans*	*Language Source*
Alutiiq	Eskimo	3	Leer 1978a, b, Fortescue et al. 2010
Dena'ina	Na-Dene	1	Kari 1977
Eyak	Na-Dene	4	Krauss 1963–1970
Tlingit	Na-Dene	2	Edwards 2009
Ts'ets'aut	Na-Dene	1	Boas and Goddard 1924
Gitksan	Tsimshianic	1	Hindle and Rigsby 1973
Haida	Isolate	0	Lawrence 1977
Haisla	North Wakashan	*6*	Haisla Nation 2020, Fortescue 2007
Coast Tsimshian	Tsimshianic	0	Dunn 1995
Nuxalk	Salish	4	Nater 1990
Heiltsuk	North Wakashan	*6*	Heiltsuk Language & Culture Mobilization Partnership 2021, Fortescue 2007
Kwak'wala	North Wakashan	2	Grubb 1977, Fortescue 2007
Nuuchahnulth	South Wakashan	3	Stonham 2005
Ditidaht	South Wakashan	2	Ditidaht Studies Program 2020
Chemakum	Chimakuan	3	Swadesh 1955b
Squamish	Salish	4	Walker 1973, Kuipers 1967
Quileute	Chimakuan	*7*	Powell and Woodruff 1976
Clallam	Salish	1	Stewart 1976
Upper Chehalis	Salish	*5*	Kinkade 1991
Alsea	Alsean	0	Frachtenberg 1920
Hanis Coos	Coosan	0	Frachtenberg 1913
Tolowa	Na-Dene	0	Anderson and Harrison 2007
Yurok	Algic	0	Robins 1958
Wiyot	Algic	0	Teeter and Nichols 1993
Hupa	Na-Dene	0	Hoopa Valley Tribe 1983
Cahto	Na-Dene	1	Anderson 2018
Yuki	Yukian	2	Sawyer and Schlichter 1984
Kashaya	Pomoan	3	Oswalt 2018
Wappo	Yukian	3	Sawyer 1965
Esselen	Isolate	2	Shaul 1995, R.F. Heizer 1952
Salinan	Isolate	0	Mason 1918
Ineseño	Chumash	3	Applegate 1992
Luiseño	Uto-Aztecan	2	Bright 1968

Source: Generated by the authors.
Note: Italicized numbers are >4.

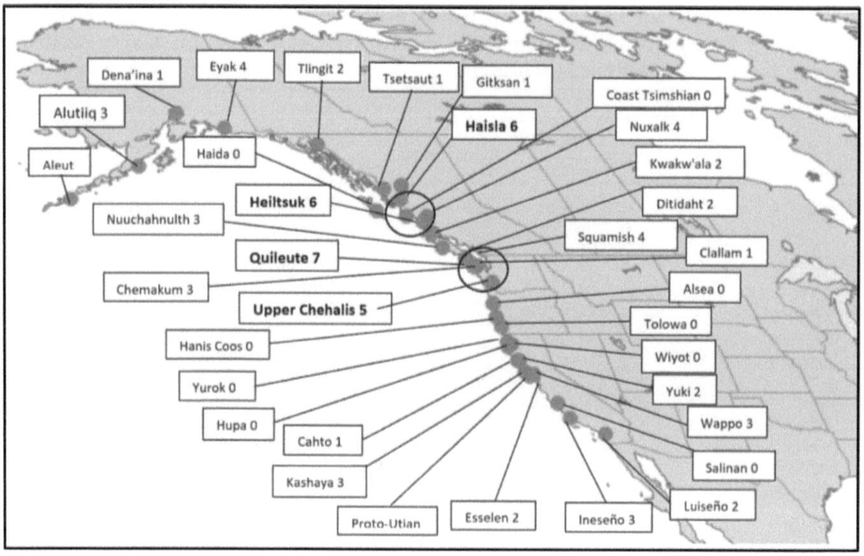

Figure 6.1 Map of the North American Pacific Coast Locating Thirty-Three Coastal Languages with Number of Potential Proto-Utian Loans Found in Each. *Note*: The two circles encompass two areas, one in British Columbia (most northern) and one in Washington State (most southern) where languages with five or more potential loans are spoken, respectively Haisla and Heiltsuk, and Quileute and Upper Chehalis. Also included are locations for Aleut and Proto-Utian (Ineseño is a member of the small Chumashan family). *Source*: Map outline adapted from Dryer and Haspelmath (2013). Permission granted by Martin Haspelmath.

potential loans from PU. The names of each of these are given in bold font and increased size in callouts, as are their respective numbers of potential loans. Two of these languages, Haisla (6 potential loans) and Heiltsuk (6) are spoken in North Coast British Columbia, and two others, Quileute (7) and Upper Chehalis (5), are spoken on or near to the Olympic Peninsula in Washington State, just south of British Columbia. Each of these areas are identified on the map by circles. These two areas are roughly 850 km apart as the crow flies.

6.4 FALSE LOANS

Some if not many of the potential Proto-Utian loanwords counted in table 6.1 are "false loans," that is, words of borrowing languages resembling Proto-Utian terms by chance. Plausibly, the more PU loans reported for a language, the greater chance these include at least one true loan rather than all being chance developments. We may then ask, "How many potential PU loanwords need be found to be reasonably confident a language has at least one true loan from Proto-Utian?" Based on results reported below, our tentative answer is at least 5. (Numbers underlined in table 6.1 are all greater than 4.)

The findings shown in table 6.2 permit a rough estimate of the number of PU loanwords expected by chance in a language not likely to have been connected (genealogically or through contact) with Proto-Utian. We follow the same four rules (section 6.2) in identifying false loans as used to identify PU loans in table 6.1. Column 1 of table 6.2 lists names of ten languages not related to Proto-Utian. These ten have in common proximity to ocean waters, and we assume their typical speakers, like those of most if not all the thirty-three languages of table 6.1, are familiar with marine ecosystems and resources. Column 2 gives the genealogical classification of each language, and column 3 its rough geographic location. Column 4 presents the number of false PU loans, i.e., words showing chance similarities to PU reflexes of PAU words. Column 5 lists sources (mostly online and hardcopy dictionaries) for the languages from which false loans have been extracted. An extension of the table follows presenting the collection of false loans found for Basque (table 6.3). Full collections of false loans for each of the other nine languages are given in online appendix D.2.

The average number of false loans for the ten languages of table 6.2 is 2.2, ranging in value from zero to four. We do not know how closely this sample (of table 6.2) approaches normal distribution, since a sample of at least one hundred languages is ordinarily required for achieving statistical significance.

Table 6.2 Number of False Proto-Utian Loans Developed by Chance in Each of Ten Coastal or Near-coastal Languages, None of Which are Related Through Kinship or Contact with Proto-Utian

Language	*Classification*	*Location*	*False PU Loans*	*Source*
Basque	(isolate)	Europe	4	Aulesria and White 1992
Irish	Indo-European	Europe	1	Ó Siochfhradha n.d.
Japanese	Japanese	Japan	3	Takahashi 1953
Korean	Korean	Korea	2	Underwood 1954
Malayalam	Dravidian	India	3	*Malayalam Dictionary (online)*
Pohnopeian	Austronesian	Oceania	0	Rehg and Sohl 1970
Suena	Binanderean	Papua New Guinea	1	Wilson and Wilson 2008
Toaripi	Eleman	Papua New Guinea	1	Brown 1968
Wolof	Niger-Congo	Senegal	2	Munroe and Gaye 1997
Zulu	Benue-Congo	South Africa	0	*Zulu Dictionary (online)*

Source: Generated by the authors.

Table 6.3 The Four False "Proto-Utian" Loans Found for Basque

Basque Word	*Proto-Utian Reflex of PAU Word*	*Rough English Gloss*
bilatu	**wel-*	"to look for"
olatu	**ʔa:len*	"wave(s)"
karkaisa	**ka:leṣ-*	"mucus, phlegm"
koskor	**kuč:i*	"small"

Source: Generated by the authors.

Eventually, we hope to develop a much larger sample to reach statistical significance but, for the present, we work with the number in hand.

Potential PU loans found for a language will always be false loans as long as the language is ascertained not to be related to Proto-Utian or to any PU relative. Even if a language were so related and four potential loans were found, these could still be chance results, given the finding for Basque (table 6.3). Hence, four potential PU loans is used as a cutoff point in our study, since at or below four *we know for certain* that all potential loans *could be false*, and above four *we do not know that for certain.*

6.5 HAISLA, HEILTSUK, QUILEUTE, AND UPPER CHEHALIS

For any language with a potential Proto-Utian loan count of four or less, our working assumption is that all of its potential loans are false (although some among them could be actual loans). Conversely, any language with a potential loan count of five or more is assumed to have at least one potential loan that is a true loan (and probably more). Under this interpretation, of the thirty-three languages of table 6.1, we assume only the four languages with five or more potential loans have at least one that is a non-chance loan, this interpretation leaving twenty-nine (or 88 percent) of the thirty-three languages all with questionable loans.

These four languages are Haisla, Heiltsuk, Quileute, and Upper Chehalis. The four languages are found on the map of figure 6.1 in two circled clusters of two languages each, respectively in North Coast British Columbia and the Olympic Peninsula region of Washington State. These are the only locations on the map of figure 6.1 in which such clustering is observed. Thus, based on this loanword evidence, extended contact of Proto-Utian speakers along the North Pacific coastal route with speakers of *many* different local languages apparently did not happen. (Perhaps of significance, the British Columbia cluster also includes Nuxalk [Salish] with four potential loans, and the Olympic Peninsula cluster also includes Squamish [Salish] with four potential loans [figure 6.1].)

The four languages are associated with three different language families: Haisla and Heiltsuk with Wakashan, Quileute with Chimakuan, and Upper Chehalis with Salish. These three families together include as members all coastal languages spoken in the region extending from the North Coast of British Columbia to just south of the Olympic Peninsula of Washington State (see map, figure 6.2). On the map, Wakashan languages are located in areas of darker shading, Salish languages by lighter shading, and Chimakuan languages by no shading at all (white, very small areas of map). The map encompasses the entire Wakashan language area, as well as the entire Chimakuan area (only two languages), but not the entire Salish language area, since the Salish languages not included are located inland (to the east, off the map). Regarding the four languages of special interest, closely related Haisla and Heiltsuk are located in the extreme northern Wakashan region of the map (figure 6.2); Quileute, one of two Chimakuan languages (with Chemakum), is located in the extreme southern region; and Upper Chehalis (Salish) is found just off the map, to the immediate south of the Clallam-language location.

Internal classifications of the three language families are presented in table 6.4. Only languages cited in the text or in maps or tables are listed. Table 6.4 also includes ASJP chronological ranges for the breakup dates of proto-languages for families and subfamilies (see Holman et al. 2011, and 9.2).

The map of figure 6.2 locates homelands for both Proto-Salish and Proto-Wakashan. Locations for these two proto-languages have been proposed by numerous scholars. McMillan (2003:248) summarizes the broadly shared understanding of where the Proto-Wakashan homeland was located: "just south of Quatsino Sound, on northwestern Vancouver Island, at the boundary between . . . Kwak'wala and Nuu-chah-Nulth languages." Kinkade (1989) discusses different Salish homeland possibilities and presents his own interpretation based on correlations between flora and fauna habitat distribution and terms for plants and animals reconstructable to Proto-Salish. Kinkade (1989:10) locates the Proto-Salish homeland as follows:

> south of the lower part of the Fraser River . . . extend[s] . . . from the Fraser River southward at least to the Skagit River, and possibly as far south as the Stillaguamish or Skykomish River. From west to east, their territory would have extended from the Strait of Georgia and Admiralty Inlet to the Cascade Mountains. A tongue of the family probably extended up the Fraser River through the Fraser Canyon.

Kinkade's (1989) Proto-Salish homeland is roughly identified on the map of figure 6.2. To our knowledge, a homeland for Proto-Chimakuan has not been discussed in the literature. Both Chimakuan languages were historically spoken close to one another on the northwest Olympic Peninsula (Washington

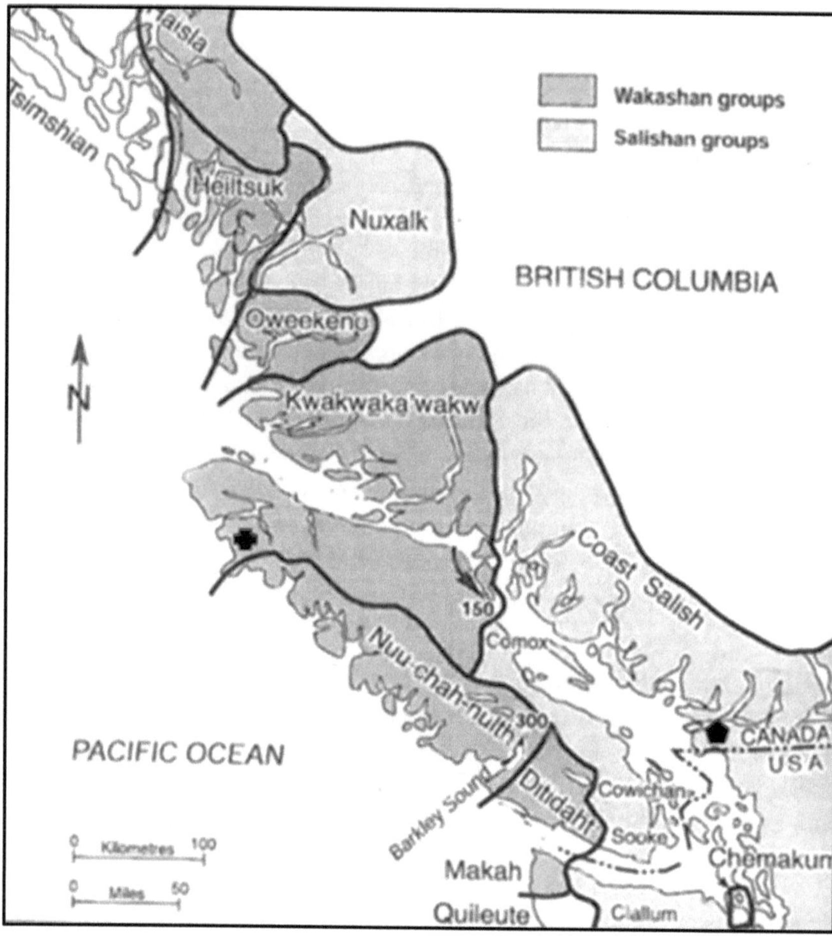

Figure 6.2 Region of British Columbia and Northwest Washington State, in which all Wakashan Languages, both Chimakuan Languages (Quileute and Chimakum), and some Salish Languages Are Located. *Note*: The *pentagon* symbol locates the Proto-Salish homeland and the *cross* symbol locates the Proto-Wakashan homeland. *Source*: Map is adapted from McMillan (2003). Reprinted with permission from the *Emerging from the Mist* by Quentin Mackie, Gary Coupland, and R.G. Matson © University of British Columbia Press 2003. All rights reserved by the publisher.

State, see figure 6.2), so the Chimakuan homeland was probably also in this same area. In any case, homelands for all three proto-languages may well have been located in the area depicted on the map of figure 6.2.

Edward Sapir was interested in these three families, believing them related to one another in a phylum dubbed "Mosan." Sapir did little research on Mosan himself, leaving this up to his student Morris Swadesh (1953a,b,c). David Beck (2000) presents more than ample evidence and rationale for

Table 6.4 Internal Classifications of Salish, Wakashan, and Chimakuan Families with ASJP Proto-Language Breakup Date Ranges (Holman et al. 2011)

Language in Classification Hierarchy			*ASJP Date Range for Proto-Language Breakup*
PROTO-SALISH			4937–2717 BP
	PROTO-COAST SALISH		(Not available)
		Nuxalk (Bella Coola)	
		PROTO-CENTRAL SALISH	3172–1746 BP*
			Quw'utsun
			Ts'ou-ke
			Squamish
			Clallam
			Upper Chehalis
PROTO-WAKASHAN			3588–1974 BP
	PROTO-SOUTH WAKASHAN		1489–819 BP*
		Nuu-chah-Nulth	
		Ditidaht	
		Makah	
	PROTO-NORTH WAKASHAN		782–430 BP*
		Haisla	
		Heiltsuk	
		Oowekyala	
		Kwak'wala	
PROTO-CHIMAKUAN			(Glottochronology breakup date: 2100 BP)
	Quileute		
	Chemakum		

Source: Generated by the authors.
*ASJP date-range based on *Ethnologue* classification, otherwise on *Glottolog* classification (9.2).
Notes: Salish Classification after Kuipers (2002); Wakashan after Fortescue (2007); and Chimakuan after Swadesh (1955b)

doubting the genealogical underpinnings of Mosan, proposing all similarities across the three groups (mostly structural) to be the consequence of diffusion within a linguistic area. Beck's understanding accords with our negative quantitative lexical findings for a Mosan language family reported in chapter 1 (online appendix A), comparisons 124, 173, 188, 199).

6.6 PROTO-UTIAN LOANWORDS IN THE FOUR LANGUAGES

Tables 6.5, 6.6, 6.7, and 6.8 give the Proto-Utian loanword inventory for each of the four languages, Haisla, Heiltsuk, Quileute, and Upper Chehalis, respectively. In each table, the first column gives Proto-Utian words borrowed by these languages. The number in parentheses following the PU word is

Table 6.5 Potential Proto-Utian Loanword Inventories for Haisla (6 Potential Loans)

Proto-Utian Word	Haisla Loan from PU	Wakashan Proto-word (Fortescue 2007)	Rough English Gloss
**pel-* (2)	*p'elh*	PNW **p'ł-*	"blink"
**pulya-* (5)	*sp'oli*	PNW **sp'uli*	"inedible red berry"
**ṣak(:)as* (21b)	*sáka sisáka*	PW **sayk*	"boat"
**li:le* (23)	*énewái*	none	"sky"
**kawisu* (34)	*ğúláli*	PW **ğawla:(li)*	"salmonberry"
**ʔo:lok* (53)	*ʔaʔewák*	none	"ocean, sea"

Source: Created by the authors.
Notes: Parenthesized numbers are IDs used for the sixty-six PAU/PU reconstructions [2.4.6]. PW = Proto-Wakashan, PNW = Proto-Northern Wakashan, ğ = voiced uvular stop.

Table 6.6 Potential Proto-Utian Loanword Inventories for Heiltsuk (Six Potential Loans)

Proto-Utian Word	Heiltsuk Loan from PU	Wakashan Proto-word (Fortescue 2007)	Rough English Gloss
**pel-* (2)	*p'łá*	PNW **p'ł-*	"blink"
**pulya-* (5)	*sp'oli*	PNW **sp'uli*	"inedible red berry"
**ṣak(:)as* (21b)	*sáka*	PW **sayk*	"boat"
**kawisu* (34)	*ğulali*	PW **ğawla:(li)*	"salmonberry"
**kole* (37)	*gḷíƛ̓ḷá*	none	"kind of grass"
**hol(:)o* (58)	*hḷúlu*	none	"butterfly"

Source: Generated by the authors.
Notes: Parenthesized numbers are IDs used for the sixty-six PAU/PU reconstructions [2.4.6]. PW = Proto-Wakashan, PNW = Proto-Northern Wakashan, ğ = voiced uvular stop.

Table 6.7 Potential Proto-Utian Loanword Inventories for Quileute (7 Potential Loans)

Proto-Utian Word	Quileute Loan from PU	Chimakuan Proto-word (Swadesh 1955b)	Rough English Gloss
**wali* (8)	*yalá*	none	"spirit"
**ṭa(:)li-ṣ* (10)	*tíʔyałi*	none	"man, male"
**li:le* (23)	*lí:la*	none	"cloud, sky"
**ku:ṭa* (35)	*ḳatsá:biłiw*	none	"clam type"
**ʔa:len* (45)	*ʔalí:wa*	none	"wave(s)"
**ʔo:lak* (50a)	*ʔàʔlótḳ* "canoe"	none	"hatch of boat, oar, row""
**molu* (59)	*bołkʷ*	none	"hair"

Source: Generated by the authors.
Note: Parenthesized numbers are IDs used for the sixty-six PAU/PU reconstructions [2.4.6].

the ID number of the Proto-Aleut-Utian comparative set to which the word belongs (chapter 2, 2.4.6). The second column gives borrowed words based on corresponding (same table row) PU words of the first column. The third column gives, if available, proto-words for which corresponding (same table row) words of the second column may be reflexes. The fourth column gives rough English glosses for the words.

Table 6.8 Potential Proto-Utian Loanword Inventories for Upper Chehalis (Five Potential Loans)

Proto-Utian Word	*Upper Chehalis Loan from PU*	*Salish Proto-Word (Kuipers 2002)*	*Rough English Gloss*
**ṭol-* (13)	*tó:l'i-*	none	"hear"
**kuč:i* (28)	*k'ec*	none	"small"
**ku:ṭa* (35)	*qʷálign*	none	"clam type"
**hol(:)o* (58)	*x̣alu*	none	"butterfly"
**mol:ɨ* (65)	*məlk'ʷ-n*	PS **məlk'ʷ*	"to cover"

Source: Generated by the authors.
Notes: Parenthesized numbers are IDs used for the sixty-six PAU/PU reconstructions [2.4.6]. PS = Proto-Salish.

In total, eighteen potential PU loanwords are found in the four languages. Five (or 28 percent) of these loans designate marine referents: clam, boat, ocean, waves, and oar. The per-language breakdown of these marine words is: Haisla (2), Heiltsuk (1), Quileute (3), Upper Chehalis (1). Quileute has been a bit more receptive among the four languages of marine words, should number of marine loanwords be indicative of the degree to which speakers of borrowing languages learned about maritime matters from Proto-Utian speakers.

Table 6.9 presents in a matrix format the number of potential PU loans shared by all possible pairs of the four languages. For example, the cell defined by intersecting Haisla (with six potential loans) and Heiltsuk (with 6) indicates these two languages share four loanwords based on the same four Proto-Utian words, these being **pel-*, **pulya-*, **ṣak(:)as*, and **kawisu* (see tables 6.5 and 6.6). Incidences of sharing could be due either to two languages of a pair each independently borrowing the same term directly from Proto-Utian, or to one language of a pair indirectly acquiring a Proto-Utian term by borrowing it from the other paired language.

Of the six language pairs of the matrix, Haisla/Heiltsuk shows the largest number of shared potential loans (with 4), while the remainder show 0–1 shared terms. With the major exception of Haisla/Heiltsuk, results reported in table 6.9 indicate sharing incidences involving either direct or indirect borrowing of Proto-Utian words have been infrequent overall. The larger number of shared items for Haisla and Heiltsuk is not surprising since these two languages are closely related both genealogically (table 6.4) and geographically (figure 6.2).

Indirect acquisition of Proto-Utian words can take place either through the type of indirect borrowing described above or through inheritance. Tables 6.10 and 6.11 present respective loan distributions of languages of Wakashan and Salish, including proto-languages from which these developments are inferred. These tables are extracted from among the eighteen charts found in online appendix D.3.

Table 6.9 Matrix of Number of Potential Proto-Utian Loans Shared by Pairs of the Four Borrowing Languages, Each Language of Which has Five or More Potential Loans

	Haisla	*Heiltsuk*	*Quileute*	*Upper Chehalis*
Haisla	----			
Heiltsuk	4	----		
Quileute	1	0	----	
Upper Chehalis	0	1	1	----

Source: Generated by the authors.

Table 6.10 Distribution of Potential Loans for "Boat" in Wakashan Languages (From Online Appendix D, D.3)

Language Group	*Source Word*	*English Gloss*
Proto-Aleut-Utian (21b)	**ṣuka-*	"boat, baidarka skirt"
Aleut	*suka-χ*	"baidarka skirt"
Proto-Utian	**ṣak(:)as*	"boat"
Wakashan	*Loan Word*	*English Gloss*
Proto-Wakashan	**sayk*	"travel (by water)"
Haisla	*sáka sisáka*	"ferry, go for boat"
Heiltsuk	*sáka*	"to use . . . a boat, car, plane . . . "
Kwak'wala	*sak-*	"use a canoe"
Nuu-chah-Nulth	*siika*	"to sail"
Ditidaht	*siikaa*	"sail (verb)"

Source: Created by the authors.

Table 6.11 Distribution of Loans for "Blink" in Salish and Wakashan Languages (From Online Appendix D, D.3)

Language Group	*Source Word*	*English Gloss*
Proto-Aleut-Utian (2)	**pɨl-*	"blink (eyes)"
Aleut	*huluχ-six* < Early Aleut **puluχ*	"to blink, wink (of eyes)"
Proto-Utian	**pel-*	"pertaining to the eyes"
Salish	*Loan Word*	*English Gloss*
Nuxalk (Bella Coola)	*p'lha*	"blink one eye"
Wakashan	*Loan Word*	*English Gloss*
Proto-North Wakashan	**p'ƛ-*	"blink"
Haisla	*p'elh*	"blink"
Heiltsuk	*p'ƛá*	"blink"
Kwak'wala	*p'elhelá*	"blinking"

Source: Generated by the authors.

Table 6.10 gives the distribution of related words for "boat" in Aleut-Utian and Wakashan languages. From this distribution, we infer that Proto-Wakashan directly borrowed the "boat" word from Proto-Utian. Subsequently, this word was inherited from Proto-Wakashan by its daughter language, Proto-North Wakashan, from which the word was passed on to historic North Wakashan languages. In this manner, Haisla, Heiltsuk, and Kwak'wala indirectly acquired the "boat" term from Proto-Utian.

Table 6.11 presents distribution of related words for "to blink" in Aleut-Utian, Salish, and Wakashan languages. Proto-North Wakashan (or, more likely, Proto-Wakashan, although not directly attested) directly borrowed the "to blink" word from Proto-Utian, passing it on to its offspring languages (thus, indirect acquisition from PU). One of the offspring, probably Heiltsuk, subsequently lent the term to Nuxalk, a Salish language ("double" indirect acquisition from PU). The close geographic proximity of the two languages (see map of figure 6.2), and the close phonological resemblance of the "to blink" words in these two languages, that is, both lacking a syllable-central vowel, respectively, Heiltsuk *p'łá* and Nuxalk *p'lha,* support the inference that Nuxalk borrowed the term directly from Heiltsuk. (The broad distribution of the "blink" word among languages of the Northwest Coast may reflect or be reflected by the prominent depiction of large, wide-open eyes in the region's well-known Native art.)

Tables 6.12 and 6.13 show a distribution of loans common among the eighteen reported in online appendix D.3 in which only one language shows a potential loan from PU, and this loan is not a constituent of a proto-language, but rather of an historic language. In table 6.12, the potential loan is Upper Chehalis (Salish) *tó:l'i-* "hear" and, in 6.13, the loan is Squamish (Salish) **yəlχ* "to look for." One or both of these words could be the only surviving reflexes of respective Proto-Salish words borrowed directly from Proto-Utian (a result of normal language attrition). As such, these words would not constitute sufficient data for reconstructing pertinent Proto-Salish words (at least two reflexes per word are needed). Another possibility is that one or both words were borrowed directly from Proto-Utian by Upper Chehalis and Squamish, respectively. If so, this potential borrowing indicates prolonged contact of Proto-Utian speakers with those of coastal Salish languages, a time period enduring from before to after the divergence of Proto-Salish into its historic daughter languages.

Table 6.12 Distribution of Potential Proto-Utian Loans for "Hear" in Salish Languages

Language Group	*Source Word*	*English Gloss*
Proto-Aleut-Utian (13)	**ṭot-*	"hear"
Aleut	*tut-*	"to hear, to get to hear"
Proto-Utian	**ṭol-*	"hear"
Salish	*Loan Word*	*English Gloss*
Upper Chehalis	*tó:l'i-*	"hear"

Source: Generated by the authors.

Table 6.13 Distribution of Potential Proto-Utian Loans for "Search For" in Salish Languages

Language Group	*Source Word*	*English Gloss*
Proto-Aleut-Utian (7)	**wel-*	"to search, look for"
Aleut	*ilγa-lix*	"to search, to look for"
Proto-Utian	**wel-*	"to search for"
Salish	*Loan Word*	*English Gloss*
Squamish	*yəlχ	"to look for"

Source: Generated by the authors.

6.7 PHONOLOGICAL EVIDENCE FOR EARLY AND LATE PROTO-UTIAN

A phonological development in the history of Aleut-Utian languages is **q* → **k*, that is, the shift of uvular segments (*q* including *ʁ*) to velar segments (*k* including *g* and *x*). When this shift occurred is unknown, but loanword evidence suggests it did not take place until after, and perhaps well after, the beginning of the maritime migration of Proto-Utian speakers from south Alaska to California. For convenience of discussion, we use "Early Proto-Utian" to refer to the PU language before the **q* → **k* sound shift and "Late Proto-Utian" to refer to the PU language after the shift was complete.

The phonological inventory of Proto-Aleut-Utian (PAU) shows both **k* and **q* segments (chapter 2, 2.4.2). The phonological inventory of Late PU is the same as that described for PU in chapter 2, showing only one of these two segments, **k*. In the development of PU from PAU, **q* has changed to **k* (PAU **q* → Late PU **k*). The entire sequence of shifts involving PAU **q* and including Early Proto-Utian is: PAU **q* → Early PU **q* → Late PU **k*. The sequence for PAU **k* is: PAU *k* → Early PU **k* → Late PU **k*.

Loanword evidence is indicative of early and late stages in the development of Proto-Utian. Some of the thirty-three languages of table 6.1 show potentially borrowed PU words with **q* segments inherited from PAU **q* that developed before Early PU **q* shifted to Late PU **k*. Others have borrowed PU words with **k* segments inherited from PAU **q* that developed after **q* shifted to **k*.

Tables 6.14 and 6.15 give examples of these loans, beginning with words with *q* borrowed from Early PU (table 6.14). One example is *qatsá:-* "butter clam," a borrowing from Early PU by Quileute, a Chimakuan language of northwest Washington State. (The full Quileute form is *qatsá:biłiw*, *-biłiw* is a suffix**)**. The PAU source for the Early PU loan is **quṭ-* "clam type" (35) which developed as **qu:ṭa* in Early PU. Quileute borrowed the word from Early PU, thus acquiring a *q* loanword (*qatsá:-*). The Early PU word then shifted to **ku:ṭa* in Late Utian, substituting **k* for **q*. No parallel change occurred for the *q* in the Quileute borrowed word and, therefore, it survived into historic times as evidence of Early PU **qu:ṭa***.** (Development of Quileute *qatsá*:- from Early PU *qu:ṭa* resulted from regressive vowel assimilation [2.4.5].)

Table 6.15 gives examples of loan words with *k* (or *g* or *x*) borrowed from Late PU by some of the thirty-three languages. One of these is *gl̓ḯƛ̓ạlá* "grass," a borrowing from Late PU by Heiltsuk. The Late PU word was **kole*, which derived from Early PU **qole*, which in turn derived from PAU **qolæ-* (but see a slightly different analysis below). Heiltsuk could have borrowed the term from Early PU instead of Late PU, in which case the loan into Heiltsuk would have been a *q* word rather than a *k* word (and, thus, would be pertinent to table 6.14 rather than table 6.15).

Loanwords analyzed in tables 6.14 and 6.15 are useful since they can suggest the relative chronology of contact events in cultural and linguistic prehistory (chapter 9). While the histories of *k* loanwords described in table 6.16 cannot be so used, they should be mentioned to avoid confusing them with the *k* loanwords in table 6.15. An example in Table 6.16 is Haisla *saka*. The source of the loan into Haisla could have been either Early PU **ṣak(:)as* or Late PU **ṣak(:)as*, both of which are traced back to PAU **ṣuka-* . Since both Early PU and Late PU words are the same, there is no way to determine from which PU source the Haisla word was borrowed.

The *k/q* distinction in loanwords is not the only evidence for Early and Late Proto-Utian. The contrast is also observed in the distribution of ejectives (glottalized consonants). Ejective consonants are usually transcribed by an apostrophic diacritic directly over or following a sound-segment symbol, for example, *p'* (a glottalized unvoiced bilabial stop). Among the array of thirty-three coastal languages of table 6.1, twenty-eight include ejectives in their

Table 6.14 Loanwords with *q* Borrowed from PU Indicating Early Proto-Utian **q* (Which Develops into Late PU **k*)

Borrowing Language	*Proto-Aleut-Utian with q (or ʁ)*	*Early PU Word with q*	*Late PU Word with k*	*Word in Borrowing Language with q*
Alutiiq	**isowiq* "harbor seal" (62)	**so:-wiq*	**so:-wik*	*isuwiq*
Eyak	**qVč-* "bone" (39) (**q'Vč-*)	**q'ič:ič*	**kɨč:ič*	*q'ahš*
Nuu-chah-Nulth	**qaw* "berry sp." (34) (**qaw'*)	**qaw'isu*	**kawisu*	*qaw'i*
Quileute	**ʔoləʁ-* "boat related" (50a)	**ʔo:laq*	**ʔo:lak*	*ʔàʔlótq*
Quileute	**quṭ-* "clam type" (35)	**qu:ṭa*	**ku:ṭa*	*qatsá:-*
Upper Chehalis	**quṭ-* "clam type" (35)	**qu:ṭa*	**ku:ṭa*	*qʷálign*

Source: Generated by the authors.
Note: Words in parentheses are revised reconstructions from those of chapter 2 (2.4.6) based on the current discussion.

Table 6.15 Loanwords with *k* (Or *g* Or *x*) Borrowed from Late PU **k* Derived from Early PU **q*

Borrowing Language	*Proto-Aleut-Utian with q (or ʁ)*	*Early PU Word with q*	*Late PU with k*	*Word in Borrowing Language with k (or g or x)*
Gitksan	**qʊn-* "bone" (36)	**qulum*	**kulum*	*sganhahl*
Haisla	**ʔɔlɔʁ-* "ocean" (53)	**ʔo:loq*	**ʔolak*	*ʔaʔewák*
Heiltsuk	**qolæ-* "grass" (37) (**qol'æ-*)	**qol'e*	**kole*	*gl̓íƛ̓la̓*
Squamish	**isowiq* "harbor seal" (62)	**so:-wiq*	**so:-wik*	*ʔasxʷ*
Clallam	**isowiq* "harbor seal" (62)	**so:-wiq*	**so:-wik*	*ʔasxʷ*
Luiseño	**qʊn-* "bone" (36)	**qulum*	**kulum*	*kulá:wu-t*

Source: Generated by the authors.
Note: Words in parentheses are revised reconstructions from those of chapter 2 (2.4.6) based on the current discussion.

Table 6.16 Loanwords with *k* Borrowed from Either Early or Late Proto-Utian

Borrowing Language	*Proto-Aleut-Utian with k*	*Early PU with k*	*Late PU with k*	*Word in Borrowing Language with k (or g)*
Eyak	**-kučɪ-* "small" (28), (**-kuč'ɪ-*)	**kuč':i*	**kuč:i*	*gut'*
Haisla	**ṣuka-* "boat" (21b)	**ṣak(:)as*	**ṣak(:)as*	*saka**
Heiltsuk	**ṣuka-* "boat" (21b)	**ṣak(:)as*	**ṣak(:)as*	*sáka**
Kwakw'ala	**ṣuka-* "boat" (21b)	**ṣak(:)as*	**ṣak(:)as*	*sak**
Nuuchahnulth	**ṣuka-* "boat" (21b)	**ṣak(:)as*	**ṣak(:)as*	*sika**
Ditidaht	**ṣuka-* "boat" (21b)	**ṣak(:)as*	**ṣak(:)as*	*siikaa**
Esselen	**kVrɔ* "foot" (29)	**kolo*	**kolo*	*kxéle*

*See discussion where these "boat" terms are interpreted as being reflexes of a Proto-Wakashan word borrowed directly from Proto-Utian (early or late stage not determined).

Source: Generated by the authors.
Note: Words in parentheses are revised reconstructions from those of chapter 2 (2.4.6) based on the current discussion.

phonological inventories. The five lacking ejectives are Alutiiq (Eskimo), Ts'ets'aut (Na-Dene), Wiyot (Algic), Esselen (isolate), and Luiseño (Uto-Aztecan). Both Proto-Utian (as reconstructed by Callaghan [2014:69]) and Aleut (an historic language) also lack ejectives.

Early Proto-Utian had glottalized consonants, while Late Proto-Utian did not. Again, loanwords represent supporting evidence for this contrast. Just as loans from PU into languages observed with *q* indicate this segment pertained to Early Proto-Utian, loans from PU into languages observed with ejectives are similarly indicative of glottalized consonants in Early Proto-Utian. The development of Late PU from Early PU involved the loss of ejectives from all words of the PU vocabulary (but see below).

Table 6.17 Loanwords with Ejectives in Wakashan, Salish, and Chimakuan Languages Borrowed from Proto-Utian Indicative of an Early Proto-Utian Stage

Borrowing Language	*Proto-Aleut-Utian with Ejective*	*Early PU with Ejective*	*Late PU with no Ejective*	*Loanword with Ejective*
Nuxalk	**pɨl*- "blink" (2) (**p'ɨl*-)	**p'el*-	**pel*-	*p'lha*
Nuxalk	**pul(y)a*- "berry type" (5) (**p'ul(y)a*-)	**p'ulya*-	**pulya*	*p'xwlht*
Upper Chehalis	*-*kučɪ*- "small" (28) (*-*k'učɪ*-)	**k'uč:i*	**kuč:i*	*k'ec*
Upper Chehalis	**ṭot*- hear (13) (**ṭot'*-)	**tó:l'i*	**tó:li*	*tó:l'i*
Haisla	**pɨl*- "blink" (2) (**p'ɨl*-)	**p'el*-	**pel*-	*p'elh*
Haisla	**pul(y)a*- "berry type" (9) (**p'ul(y)a*-)	**p'ulya*-	**pulya*	*sp'oli*
Heiltsuk	**pɨl*- "blink" (2) (**p'ɨl*-)	**p'el*-	**pel*-	*p'łá*
Heiltsuk	**pul(y)a*- "berry type" (5) (**p'ul(y)a*-)	**p'ulya*-	**pulya*	*sp'úli*
Heiltsuk	**qolæ*- "grass" (37) (**qol'æ*-)	**qol'e*	**kol'e*	*gḷ́íƛ̓ḷá*
Kwak'wala	**pɨl*- "blink" (2) (**p'ɨl*-)	**p'el*-	**pel*-	*p'elhelá*
Nuu-chah-Nulth	**qawV*- "berry" (34) (**qaw'V*-)	**qaw'isu*	**kawisu*	*qaw'i*
Chemakum	*-*kučɪ*- "small" (28) (*-*k'uč'ɪ*-)	**k'uč':i*	**k'uč':i*	**k'at'in*

Source: Generated by the authors.
Note: Words in parentheses are PAU reconstructions revised here from 2.4.6.

Table 6.17 gives words with ejectives in Wakashan, Salish, and Chimakuan languages borrowed from Early Proto-Utian. Consider the Haisla word for "to blink," *p'elh*. This word was directly borrowed from Early PU **p'el*-, which subsequently lost *p* glottalization with the shift from Early to Late PU, becoming **pel*-. All these words trace back to PAU **pɨl*-, which, because of loanword phonology, we now know reconstructs with stem-initial **p'* rather than with stem-initial *p* (see the replacement PAU reconstruction, **p'ɨl*, given in parentheses in the table).

Two loanwords cited in table 6.17 are doubly indicative of Proto-Utian developmental stages. Both the *q* and *w'* in Nuu-chah-Nulth *qaw'i* "berry" indicate borrowing from Early Proto-Utian (i.e., **qaw'i*-). The other is Heiltsuk *gḷ́íƛ̓ḷá* "grass." In this case, *g* is indicative of Late Proto-Utian and *ḷ'* is indicative of Early Proto-Utian, suggesting the loanword was acquired from a transitional stage between Early and Late Proto-Utian.

Chapter 7

Ethnobiological Setting

7.1 INTRODUCTION

This chapter provides environmental background on the Utian coastal journey to California from Alaska in the Middle Holocene (9.2), focusing on the different ecosystems these travelers might have experienced along the way. We also consider past climate and the role of ancient sea levels that influenced peoples' lives, but which also constrain our ability to find evidence of those lives. We briefly touch on potential connections with other Indigenous Groups the travelers may have encountered (chapter 6), whose knowledge, practices, and languages may have influenced their own knowledge systems, and vice versa. By drawing together fragmentary information from the fields of archaeology, ethnoecology, biogeography, paleobotany, and paleozoology, we set the stage for a previously unrecognized relationship between Aleut and Utian peoples.

The ecological setting outlined here for the Middle Holocene focuses on plant communities and is based on plant distributions reflected in the palynological (pollen) record, as well as on knowledge of plant distributions along the coast today and how they relate to historical ranges of these species. Our synthesis is spotty in places, even for these data, since it depends on the focus of other researchers' efforts.

We are on even less firm ground when reconstructing the animal communities because the paleo-zoological record for this period is generally sparse. Nevertheless, based on available zooarchaeological information, people of the Early Holocene seem to have exploited the same broad suite of terrestrial and marine resources—shellfish, fish, birds, and marine and land mammals—as in the Middle Holocene, with emphasis varying in time and place (see information from Hebda and Frederick's [1990] archaeological summary of

Northeast Pacific marine animal resources since the Pleistocene, and Moss et al. [2007] overview of Middle Holocene resources, as well as zooarchaeological studies from other time periods in places such as Haida Gwaii [e.g., Cohen 2014], and Fedje et al. [2005] for Pleistocene/Early Holocene data from Kiilgii Gwaii).

From linguistics, some of the animal names reconstructable to Proto-Aleut-Utian, for example, "sea lion," "sea otter," "clam," and "coho salmon" (chapter 9, 9.4.1), indicate that representatives of these species or broader taxa were encountered by Utian travelers on their coastal journey. The longer the duration of their journey, the greater the likelihood they encountered new plant and animal species along the way, and the longer their stays in various locales, the greater the chance they exchanged words with speakers of other languages (chapter 6).

7.2 TIMING OF THE JOURNEY

The Middle Holocene was a dynamic period on the North American coast. Modern ecological communities of the region were being established (Hebda 1995) following the glacial retreat of the Pleistocene era and the rapidly shifting climate of the Early Holocene. These changes were especially dramatic in previously ice-covered regions. From Alaska to northern Washington State, the Early to Middle Holocene was a time of relatively warm climate and gradually evolving ecosystems, when many coastal species we see today began to occupy their modern ranges. During this dynamic time, peoples of the region settled in and increasingly learned how to interact with the plants and animals around them. One aspect of these adaptations was the development of social systems allowing people to live with and from the resources and environments they encountered.

One of the more thoroughly investigated regions of our area of interest is coastal British Columbia (BC). Prior to the Middle Holocene, much of the landscape on the BC coast was covered in coniferous forest or woodland, first with shore pine (*Pinus contorta*), and then, by the start of the Holocene around 11,700 years ago, western hemlock (*Tsuga heterophylla*) and Sitka spruce (*Picea sitchensis*), as well as shore pine. The period between 11,650 and 10,000 years ago is considered the warmest time of the entire Holocene on the northwest coast of North America (the early climatic optimum).

The Middle Holocene climate of the region was cooler and moister than that of the Early Holocene, although warmer and somewhat drier than today. Ecosystems across the region had adjusted to the new, warmer temperature regimes in the Early Holocene, and the modern-day coastal temperate rainforests began to establish themselves in areas that had once been mixed conifer forests (Hebda 1995; Brown et al. 2022). In some places, such as southeastern Vancouver Island, grasslands flourished during the warmest, driest times but began to evolve into conifer forests before the Middle Holocene (Hebda 1995). While specific data from the archaeological or paleozoological records for this period are sparse, animal populations would have shifted with changing climates.

Sea-level histories in the Middle Holocene are variable and region-specific, depending on local tectonics and whether the region was covered in ice (James et al. 2009, Lepofsky et al. 2024). The impacts of changing climate and coastlines on animal life and their subsequent effect on the peoples who used these species for food and other purposes were undoubtedly profound but would have been highly variable and are not well-documented (Moss et al. 2007).

Wetlands began to expand and change in the Middle Holocene in response to cooling and moistening climates (Brown and Hebda 2003), and the geographic range of important forest species, such as western red-cedar (*Thuja plicata*), yellow-cedar (*Callitropsis nootkatensis*) and possibly Pacific yew (*Taxus brevifolia*), was beginning to move northward. Western red-cedar, so important for Northwest Coast cultures in more recent millennia, was slower to establish than some of the other rainforest species, although its seeds were found in the early Kiilgii Gwaii site on Haida Gwaii (Cohen 2014) and leaf scales on Brooks Peninsula before the Middle Holocene (Hebda 1997). Western red-cedar only became a major forest tree species toward the end of the Middle Holocene (Hebda and Mathewes 1982), however, not having been a dominant part of the plant assemblage in northern BC and Alaska until then (Lacourse et al. 2012). Other culturally valued trees and shrubs, such as pines (*Pinus* spp.), alders (*Alnus* spp.), western hemlock, and berries (e.g., *Rubus* spp., *Vaccinium* spp.), were likely already well-established more widely, but their ranges shifted, in some cases expanding in response to cooling and moistening climates in the Middle Holocene. Several of these species may have spread from unglaciated refugia along the coast considerably earlier (C. Hebda et al. 2022). Certainly, a greater diversity and abundance of species was available to the south of the glacial ice limit than northward.

7.3 PALEO-CLIMATE AND SEA-LEVEL CHANGES

In this section, we describe briefly what is known about past climatic conditions and relative sea-level histories of the greater Kodiak region of Alaska, parts of the Northwest Coast, and the San Francisco Bay Area (SFBA) during the late Early Holocene and the Middle Holocene. Each of these regions is unique ecologically and experienced its own trajectory of geomorphologic processes. These physical processes not only influenced how people related to each other and to their surroundings but also impacted our ability to find evidence of these past relationships in the archaeological record.

7.3.1 The Greater Kodiak Region of Alaska

The late Early Holocene in southeast Alaska was a time of dramatic ecological and geomorphic changes. At this time, sea levels dropped or rose dramatically, depending on whether the region was glaciated during the Ice Age (Baichtal and Carlson 2010; Baichtal et al. 2021). By the Middle Holocene, sea levels had approached current conditions but continued to be variable due to ongoing isostatic and tectonic shifts. Finding archaeological sites in coastal regions dating to this time period has been hampered by these dramatic landscape changes. For instance, on Kodiak Island, while the earliest sites (the "Ocean Bay" sites; see chapter 8, 8.4.1) correspond in time with the stabilization of sea levels, people were possibly on the islands earlier (Patrick Saltonstall, pers. com. 2023).

After a period of cold winters and warm summers in southeast Alaska, the climate cooled beginning about 6,800 years ago (Baichtal et al. 2021). In the Kodiak region specifically, the pollen and macrofossil records suggest that the climate was relatively warm and moist until *c.* 6,500 years ago, after which it became considerably drier, at least during the growing season. This shift may be associated with a trend toward colder temperatures in the Kodiak region, but neighboring regions in Alaska remained warmer during this period (Peteet et al. 2019). For the Arctic, the regional record of glacial advances, as well as other climate proxies, supports the interpretation of cooling in most northern regions beginning *c.* 5,000 years ago (Kaufman et al. 2016; Kjaer et al. 2022), culminating in what paleoecologists call the Neoglacial. The first people to inhabit the Kodiak Archipelago arrived during the period of relatively warm and moist climates, at least 7,500 years ago.

Not surprisingly, it is difficult to develop hypotheses regarding the effects of climatic patterns on peoples' lives from this complex Holocene paleoclimatic record. If, for instance, the climate shifts made it harder for travel to important resources, or made resources more unpredictable, significant changes to the subsistence system would have occurred. Colder weather alone, however, may not have led to such changes. Regional culture history expert Amy Steffian posits that adapting to stormier weather over the long term may have contributed to the eventual diversification and intensification of subsistence systems some 4,000 years ago (Steffian, pers. com.; Steffian et al. 2006). Notably, around 800 years ago, Alutiiq (Yupik Eskimo) people responded to the further cooling of regional temperatures by moving their village to access salmon streams and to hunt more whales (title page,

Alutiiq Museum and Archaeological Repository website, 2022). Thus, cooler weather associated with the Neoglacial of the Middle Holocene may have also motivated population movements and was perhaps the impetus for the Utian southward migration.

7.3.2 The Northwest Coast

Changes in relative sea level associated with melting glaciers had profound and region-specific effects on the landscapes of the Northwest Coast in the Middle Holocene (e.g., Shugar et al. 2014). In Heiltsuk territory, sea-level changes were relatively small throughout the Holocene because of the position and thickness of the ice. Currently, there is no sea-level record for Haisla traditional territory; however, work in the neighboring Tsimshian territory to the west indicates that the previously much higher sea-level stand approached current conditions after *c.* 5,500 years ago and possibly as late as 2,000 years ago (Letham et al. 2021; Mackie et al. 2011).

By the early Middle Holocene, sea level had stabilized in many areas of the coast (Mackie et al. 2011). This stability allowed people in those regions to settle more fully, such that most places on the coast supported significant Indigenous occupations by the Middle Holocene (e.g., Lepofsky et al. 2024; McLaren et al. 2015). It was into this well-established social and ecological setting that Utians would have entered on their travels southward.

The territories of the Quileute and Upper Chehalis, the two more southerly groups whose languages contain loans from Proto-Utian, are beyond the southern limits of the glacial ice. Such coastal areas would have been attractive places for early settlers to live; however, global sea-level rise (caused by the melting of the ice sheets), combined with tectonic activity, has meant that the Early and early Middle Holocene coastal archaeological record is relatively unknown (e.g., Losey 2003). As a result, the presence of early occupations is largely speculative (e.g., Croes and Kucera 2017). By *c.* 5,500 years ago, however, the flooding of attractive living places, such as estuaries, had slowed considerably. Consequently, in many regions, the extant archaeological record begins after that time (Moss et al. 2007).

The Middle Holocene along the central and southern Northwest Coast witnessed similar changes in climate to those in coastal Alaska, but with different timing. The onset of the Neoglacial, when temperatures were relatively cooler, is generally thought to be about 5000 BP (e.g., Walker and Pellatt 2003). This timing is by no means universal, however, and cooler temperatures may have come later in more southerly regions (Moss et al. 2007). Data are limited, but a novel paleo-ecological method suggests that this change in climate was accompanied by relatively cooler sea surface temperatures, which, in turn, influenced the fish that people harvested (Hillis et al. 2022).

Undoubtedly, environmental changes associated with the Neoglacial would have influenced past peoples' lives in a myriad of ways (e.g., Moss et al. 2007). However, linking climate change to changes in culture often requires ecological and cultural details that are not readily available in the paleo-records (cf. Lepofsky et al. 2005).

7.3.3 The San Francisco Bay Area

Post-glacial changes in sea level also played a dramatic role in landscape change in the greater SFBA in the Middle Holocene. San Francisco Bay is a drowned river valley created relatively recently by rising sea levels in post-Pleistocene times. Geological studies of sediment cores (e.g., Atwater et al. 1977, 1979; Meyer and Kaijankoski 2017) indicate that it began to fill quickly about 10000–9500 BP and continued to increase rapidly in size until about 6000–5000 BP, when sea levels began to stabilize. At this later time, the broad outlines of the bay as we know it today began to emerge. Geological studies indicate the rate of bay infilling was so rapid (on the order of water advancing up to 30 meters a year) that it would have been difficult for people to settle permanently in specific places along its shore until the rate of increase eased off, probably at the earliest around 6000–5000 BP. After this time, the modern configuration of the San Francisco estuary system began to take shape. It offered exceptional opportunities for new settlements, including access to biologically rich intertidal flats, wetlands, and riparian zones. Thus, for Utian people journeying from Alaska to California and for the already resident peoples of the region, the newly created Middle Holocene landscape of the SFBA offered a potentially rich, protected place to call home about 6,000 years ago.

7.4 OVERVIEW OF HUMAN–COAST RELATIONSHIPS

The number of people living on the Northwest Coast by the beginning of the Middle Holocene is not known. Thanks to increasingly refined field methods for finding Late Pleistocene and Early Holocene archaeological sites, we know that people were living throughout the coastal area at least by 14000 BP (Lepofsky et al. 2024; McLaren et al. 2018; Fedje et al. 2005; A. Mackie et al. 2003; Q. Mackie et al. 2011; Moss 2011; Wigen 2005; R.L. Carlson 1996). Putting faces on our imaginings of these early people is challenging, but the recent discovery within the present territory of the Heiltsuk (Haíłzaqv) Nation of twenty-nine 13,000-year-old human footprints helps to visualize them as living people. The footprints, embedded in clay layers, are impressions of two adults and a child who walked together on the sands of Calvert Island at that time (McLaren et al. 2018). While there is a long-standing debate about how people came to the region from Beringia (the open land area between Alaska and East Siberia during the Pleistocene), it is widely accepted that the Calvert Island footprints were made by sea-going people who traveled down the coast following glacial refugia (Q. Mackie et al. 2011) and that the route, termed the "Kelp Highway" by some, was used by ancient mariners (Fladmark 1979; Erlandson et al. 2007; Erlandson, Moss, and Des Lauriers 2008; Davis and Madsen 2020; Braje et al. 2020). Early sites in places like Haida Gwaii reflect the time depth of people's marine and terrestrial associations (Fedje and Mackie 2005).

In large part because of post-glacial changes in the landscape (sea-level changes, deposition, erosion, etc.), archaeological finds for this period are sparse (Moss et al. 2007), and for the most part, we know little about how coastal peoples lived, what their settlements looked like, or how they may have interacted with each other. Based on the scant information available, we know the Middle Holocene was a time when many peoples were becoming increasingly connected to specific territories, as evidenced in sites situated in less geomorphologically dynamic locations showing consistent occupation in specific locales for millennia from the Early Holocene onward (McLaren et al. 2018). In places where sea-level changes were too dynamic for settling in for multiple generations, we see people connecting to larger landscapes. By the time of the Utian migration from Alaska to California, these travelers would have encountered diverse groups of people who were well connected to their specific homelands. Prior to this time, in the Late Pleistocene and Early Holocene, peoples living along the coast may have been less sedentary, with more temporary settlements and gathering places, relying on local resources and traveling to new resource sites as these were exhausted.

By that time, coastal peoples would have already developed strong ties with their local forest and ocean environments, with distinctive, locally situated foods, technologies, languages, and other cultural traits. Many of the folk stories of contemporary Native Americans living along the coast reflect continuous residency in their homelands "since the beginning of time." However, in some cases, their stories describe tumultuous episodes from the past, in which floods, volcanic eruptions, earthquakes, tsunamis, and major climatic shifts required migrations. Other stories describe individuals arriving from unknown areas and settling in certain locales, sometimes joining existing villages (e.g., Harris 1901; Cove and MacDonald 1987; Boas 2002; Cruikshank 2005; Ludwin et al. 2005; Ives 2010; Atleo 2011). Added to other evidence, such oral histories suggest periods of travel and resettlement along the Northwest Coast and southwards (10.4.3.1).

Given the high variability in environments and cultures across the Middle Holocene, and from north to south along the Northwest Coast route, we can only speculate about the numbers of people and ecosystems the Utians would have encountered in their travels. We can, however, assume the environments they experienced along the way would have varied substantially from north to south, and that the peoples they met would have informed and shaped their own culture in ways reflected in their vocabulary and possibly in other traits yet to be identified.

Based on known congruities and similarities in technologies, foods and materials, and oral traditions (Turner 2014), Middle Holocene coastal peoples would likely have continuously borrowed and exchanged technologies, products, and names for things with their neighbors, both coastal and inland, and with others passing through. They would have traveled to and camped in different locales within their territories, such as the outer coast, inland river valleys, and possibly even montane regions, as seasonally different situations required. They would also have been adept not only at harvesting a range of foods, from diverse shellfish and fish to marine and terrestrial mammals, birds and birds' eggs, berries, and edible bulbs and roots, but also processing these foods for winter storage and trade. Trade networks would already have been well developed, even in the Early Holocene (Carlson 1994). The knowledge for various management practices, such as landscape burning, selective harvesting, weeding, pruning, reseeding, tilling, transplanting, and terracing (Anderson 2005; Deur and Turner 2005; Lightfoot and Lopez 2013; Lightfoot et al. 2021), which appear in the archaeological record as being well developed by at least 4,000 years ago (Grier 2014; Hoffman et al. 2016; Lepofsky et al. 2024; Lepofsky and Armstrong 2018; Toniello et al. 2019), must have had their origins during the Middle Holocene. The presence of fish traps, for salmon in this case (Moss et al. 2007), from the archaeological record of around 6,000 years ago at Namu on the central BC coast, is a clear indication of management systems from that period.

In chapter 6, based on linguistic evidence of words borrowed from Proto-Utian into coastal languages, we propose migrants experienced at least two prolonged stops after leaving Alaska, one in the north coastal region of British Columbia, and one on the Olympic Peninsula of Washington State, before reaching coastal central California. We describe here some potential environmentally and culturally salient features Utian travelers might have encountered in these different regions, at one point with North Wakashan language speakers and at another with Chimakuan and Salishan speakers on the Olympic Peninsula, who may well have influenced them culturally and linguistically, and whom they, in turn, may also have influenced in some ways. We focus on plant species known to be culturally valued for food, material, and other purposes. Current distribution maps from E-Flora BC (Klinkenberg 2023) have helped determine potential past ranges of certain species along the coast, as have several paleo-ecological studies (e.g., Brown et al. 2008; C. Hebda et al., 2022).

7.5 COASTAL SOUTH ALASKA

When Utians launched their journey to the south, the environment of coastal south Alaska in the vicinity of the Kodiak Archipelago, where it probably began, would have been similar in some ways to that experienced by people in recent centuries, with some of the plants they encountered similar to those existing there today. Moss et al. (2007) describe the Middle Holocene climate of southeastern coastal Alaska as variable, but with warmer, wetter conditions than earlier. Based on pollen analysis at a site on Kodiak Island, Peteet et al. (2019) observed that peatlands with *Sphagnum* moss generally diminished through the Middle Holocene, and woodlands of alder (*Alnus* sp.) and Ericales (probably crowberry, *Empetrum nigrum*) increased. Depending on where these early people lived, they would also have been familiar with western hemlock, shore pine, Sitka spruce, and willows (*Salix* spp.). They were also likely familiar with bears, sea lions, dolphins, seals, gulls, ducks, rockfish, salmon, clams, and other northern animal life (see Moss et al. 2007). Their northern homeland would have been significantly less diverse, however, than the biological richness they encountered further south along the coast.

7.6 NORTHERN NORTHWEST COAST

Utians were in contact with ancestral Northern Wakashan peoples (Haisla and Heiltsuk [Haíłzaqv]) on the central coast of what is now British Columbia and introduced words into these languages (chapter 6, 6.3). Traveling through this region, Utians would have encountered a developing form of the coastal temperate rainforest present in the region today, consisting of Sitka spruce, western hemlock, lodgepole/shore pine, and possibly cedars and alder. They would also have encountered plants and animals of many habitats, including rocky and sandy shorelines, coastal estuaries, peat bogs, and alluvial forests of cottonwood (*Populus balsamifera*), willows, and alders. All these species provide materials and medicines, and the inner bark and cambium of all of them, except perhaps for willows, are edible and may have provided nutritious food in the late spring. Depending on the timing of their travels, Utians likely encountered yellow-cedar and possibly even western red-cedar, as these iconic trees of immense recent cultural importance expanded their ranges during the Middle Holocene, although they may not have been as abundant or as large as in historic and Late Holocene times (Hebda and Mathewes 1982). For Utians, they would have been sources of easy-to-work wood for building canoes, repairing their boats, and for implements, as well as of fibrous bark, roots, and branches for basketry, clothing, hats, fish traps, and other items.

Douglas-fir (*Pseudotsuga menziesii*) grew on the drier side of Vancouver Island and the Gulf Islands (K.J. Brown et al. 2008), beginning in the Early Holocene. This tree species was widespread in the Middle Holocene, and its range extended southward to central California (Moss et al. 2007), but it did not occur on the north and central North Coast of British Columbia. Although the tough-wooded Pacific yew currently grows as far north as Haida Gwaii and southeastern Alaska and its remains have been identified from an Early Holocene Vancouver Island site (Fitton and R. Hebda 2004), it is not common today on the central coast of British Columbia and may not have been seen there by Utian travelers.

One fruit species, soapberry (*Shepherdia canadensis*), has certainly been present in coastal areas since the late Pleistocene (Mathewes 1973). Some fruit species, such as Pacific crabapple (*Malus fusca*), highbush cranberry (*Viburnum edule*), some kinds of blueberries, huckleberries, and cranberries (*Vaccinium* spp.), salmonberry (*Rubus spectabilis*) and thimbleberry (*R. parviflorus*), salal (*Gaultheria shallon*), red elderberry (*Sambucus racemosa*) gooseberries and currants (*Ribes* spp.), saskatoonberry (*Amelanchier alnifolia*), kinnikinnick (*Arctostaphylos uva-ursi*), and seaside strawberry (*Fragaria chiloensis*), were also likely present in early North Wakashan territory and therefore would have been available to migrating Utians. (Genetic evidence suggests earlier peoples may have transported Pacific crabapple with them from Asia [Routson et al. 2012].) The iconic medicinal plant devil's-club (*Oplopanax horridus*) was also likely encountered all along the northern portion of the coast. All these species occur today in Alaska and are common along the Northwest Coast. They are well-known to all present-day Native Northwest Coast groups, indicating a deep history of human use. Proto-Utian names for salmonberry and for trailing currant (*Ribes laxiflorum*) or one of its relatives were possibly borrowed into North Wakashan (see chapter 6).

Other plants with extensive northern ranges and cultural recognition likely encountered by Utians in this area include the edible green vegetables cow-parsnip (*Heracleum maximum*) and western dock (*Rumex fenestrata*); stinging nettle (*Urtica dioica*), whose stem fibers are a cordage and fish net material; and the root vegetables spiny wood fern (*Dryopteris expansa*), northern riceroot (*Fritillaria camschatcensis*), Pacific hemlock parsley (*Conioselinum pacificum*), and Pacific silverweed (*Potentilla egedii*) (see Compton 1993a, b). Cow-parsnip, a key food plant, has occurred on the coast since late Pleistocene times (Hebda et al. 2005) and doubtless occurred widely in the Middle Holocene.

Marine plants and algae would also have been among those species seen by and possibly already known to Utians from their Alaskan homelands: eelgrass (*Zostera marina*), whose rhizomes are edible; red laver seaweed (*Pyropia abbottiae* and related species), giant kelp (*Macrocystis integrifolia*) and bull kelp (*Nereocystis luetkeana*), among others. All these marine species have long been known to and used by First Peoples from Alaska and the Northwest Coast, south to California, for food, materials, medicines, and other purposes (Kuhnlein and Turner 2020; Turner 2014) and may have been important resources for Utians all along the travel route. Moreover, extensive kelp beds all along the coast provided rich habitats for diverse fish, shellfish, and marine mammals, as well as shelter and anchorage for the travelers throughout their journey.

7.7 OLYMPIC PENINSULA

The second proposed region of significant Utian contact with coastal peoples along their route to California is on the northern Washington coast, on the western side of the Olympic Peninsula, homeland of the Quileute, the coastal group with the most vocabulary borrowed from Proto-Utian (at least seven terms [chapter 6, table 6.4].) Upper Chehalis (Salish), whose speakers occupy territory today south and inland from the Quileute area, also received at least five loans from Proto-Utian (table 6.4). Quileute and Upper Chehalis territories are in coastal rainforest vegetation, with west-coast conifers, alders, and other deciduous trees, and a dense undergrowth of shrubs. Peat bogs and grassy prairies also likely provided their own distinctive vegetation, like that possibly encountered further north.

As well as the tree, shrub, and herb species mentioned earlier as likely to have been encountered by Utians in the North Wakashan area, these travelers likely came across some new species, including Douglas-fir (mentioned above), grand fir (*Abies grandis*), western white pine (*Pinus monticola*), and more frequently Pacific yew, although, as noted previously, they may have already encountered the latter species further north (Wainman and Mathewes 1987; Fitton and R. Hebda 2004). If they traveled high into the mountains, they might have also encountered subalpine fir (*Abies lasiocarpa*).

New broad-leaved tree species encountered on the Olympic Peninsula would potentially have been bitter cherry (*Prunus emarginata*), Pacific flowering dogwood (*Cornus nuttallii*), and three maple species (*Acer macrophyllum, A. glabrum,* and *A. circinatum*). Arbutus or madrone (*Arbutus menziesii*) and Garry oak (*Quercus garryana*), trees associated with drier lowland areas from Vancouver Island southwards, may not have been encountered on the west coast of the Olympic Peninsula in the Middle Holocene. New shrubs might have included evergreen huckleberry (*Vaccinium ovatum*), oceanspray (*Holodiscus discolor*) (although this species occurs today in the Bella Coola Valley of the central BC coast), trailing blackberry (*Rubus ursinus*), Oregon grape (*Mahonia aquifolium*), and possibly hazelnut (*Corylus cornuta*), although its distribution on the Olympic Peninsula in more recent times tends to be further inland. Potentially new herbaceous species include wapato (*Sagittaria latifolia*), cattail (*Typha latifolia*), as well as tule (*Schoenoplectus* spp.), the stalks of which Native Americans used to make boats in California and elsewhere.

7.8 CENTRAL CALIFORNIA COAST

As Utians journeyed further south, stopping at other locales *en route*, they would have encountered a warmer, drier, though still somewhat variable climate (Palmer et al. 2023). They would not find some of the northern species familiar to them, although others, like cottonwood, salal, oceanspray, wild rose, red huckleberry (*Vaccinium parvifolium*), cow-parsnip, and seaside strawberry, would still be found. The more cold-tolerant northern species, however, would have given way to warmer climate species growing along the coasts of Oregon and California, including those available to historic Miwok and Costanoan peoples (descendants of Proto-Utians) (Beckwith 1995; Stevens 2004).

Trees ancestral Utians were likely to encounter, maybe for the first time, on the California leg of their journey include arbutus or madrone (*Arbutus menziesii*), and Garry oak, as well as other oak species (*Quercus kelloggii, Q. chrysolepis, Quercus agrifolia*, and *Q. densiflora*). New conifers would likely have included sugar pine (*Pinus lambertiana*), Port Orford cedar (*Chamaecyparis lawsoniana*), and coast redwood (*Sequoia sempervirens*). Shrubs perhaps not encountered previously in their travels may have included blue elderberry (*Sambucus cerulea*), hazelnut (*Corylus cornuta*), mock orange (*Philadelphus lewisii*) and manzanita (*Arctostaphylos* spp.). Sandbar willow (*Salix exigua*) and redbud (*Cercis occidentalis*) are two fiber-producing species that they would have seen and learned about at their California destination.

Herbaceous species potentially new to Utians include those with edible underground parts: wapato (*Sagittaria latifolia*), camas (*Camassia* spp.), chocolate lily (*Fritillaria affinis*), nodding onion (*Allium cernuum*) and other wild onion species, wild caraway (*Perideridia gairdneri*), and springbank clover (*Trifolium wormskioldii*), all species found on the southern BC coast today. Two likely others are mariposa lily (*Calochortus* spp.) and blue dicks (*Dichelostemma pulchellum*). Other species of coastal California of cultural importance are hemp dogbane (*Apocynum cannabinum*), tobacco (*Nicotiana quadrivalvis*), and coastal mugwort (*Artemisia douglasiana*), a medicinal and ceremonial species. Notably, some of these species may have been transported northwards along the Northwest Coast by travelers in prior centuries or millennia (Turner et al. 2021). All these species are known to and named by more recent descendants of Utian voyagers and represented valuable resources for them when they first arrived.

At some point in their journey, possibly shortly after their arrival in the San Francisco Bay Area and subsequent wider dispersal (chapter 9, 9.4.4), Utian migrants acquired expertise in harvesting and processing acorns, including grinding and leaching techniques to remove some of the bitter, potentially toxic tannins. They also began using a diversity of protein- and fat-rich seeds such as goosefoot (*Chenopodium* spp.) as a major source of nutrition (Anderson 2005; Milliken et al. 2007; Wohlgemuth 2004:144–145). They may have learned some of these processing practices from local residents, since a strong presence of grinding tools (milling slabs and, later, mortars and pestles) in the archaeological record of the Early and Middle Holocene is indicative of acorn processing and seed grinding. Mortars were in use at least 5,700 years ago in interior locations of the greater SFBA some distance away from the expanding bay system (Byrd et al. 2017:10–6; Rosenthal and Meyer 2004a:34–35; Wohlgemuth 2004:143; Milliken et al. 2007).

Although they would already have been familiar with some carbohydrate-rich root vegetables such as spiny wood fern (*Dryopteris expansa*), northern riceroot (*Fritillaria camschatcencis*), and Pacific silverweed (*Potentilla egedii*), the plethora of other species (named previously) to be found in California would have markedly enriched their dietary diversity. Specifically, the processing and use of inulin-containing foods like camas (*Camassia* spp.), blue dicks (*Dichelostemma* spp.), and onions (*Allium* spp.) gave their descendants wide recognition as geophyte experts (Anderson 2005).

As we have noted, the record of animal distribution at the time of Utian travels along the coast and that of the potential use of different animal species for food and other purposes is not well-documented (McKechnie and Wiggen 2011; but see Hebda and Frederick 1990 for potential examples). As with new plants encountered along their route, Utians would have become acquainted with several new species of animals, such as the California sea lion (*Zalophus californianus)* (replacing the Steller sea lion [*Eumetopias jubatus*] of northern climes) and California red abalone (*Haliotis rufescens*), as they journeyed south. In general, however, in spite of their comparative mobility, in some cases migrating considerable distances annually, the faunal populations would likely have been more consistent along the coast, despite local and regional variation in the numbers of at least some species.

In sum, the linguistic evidence of commonalities in vocabulary, as well as genetic evidence and similarities in many cultural traits, all indicate that encounters of Utians with other Indigenous Groups on their southward journey resulted in a significant exchange of cultural knowledge about plants, animals, and environments.

Chapter 8

Archaeological Perspective

8.1 INTRODUCTION

Combining linguistics and archaeology to further our understanding of the migration of cultural groups in the distant past has a long tradition. On the one hand, linguists turn to archaeologists to determine within a narrow timeframe when people speaking certain languages moved to new locations. On the other, archaeologists look to linguists to associate a specific cultural group with the otherwise "faceless" collection of artifacts and features making up the archaeological record. While the combination of data from the two disciplines into "linguistic prehistory" or "linguistic archaeology" is potentially a powerful way to understand past lives more fully and respectfully, the literature provides many examples where these two datasets have been combined incautiously to produce stories that do not fit comfortably with either the original archaeological or linguistic data (see Adams et al. 1978, Hughes 1992).

Within archaeology itself, scholars and Indigenous communities today struggle with the attribution of changes in the archaeological record to the in-movement of non-local peoples. Many such examples from the archaeological literature of the early 1900s attribute major changes in the archaeological record to the in-migration of outsiders rather than in situ cultural shifts (see Adams et al. 1978:489–496). By the 1950s, however, "fascinating accounts which result from the oft-proposed but seldom documented movements of prehistoric people" were beginning to be questioned for their naivety and lack of supporting evidence (Thompson 1958:v). As a result, beginning in the 1960s and 1970s, archaeologists began to temper their use of migration models to explain changes in the archaeological record (Adams et al. 1978, Heggarty 2014:599, Mallory 2021:273, Rouse 1986:xii). The disenchantment with migration models stems from many factors, including the *ad hoc* way in which many interpretations were generated, the lack of consideration of other possible explanations such as diffusion, climate change, and sociopolitical developments, the dearth of supporting evidence for population movements, and the paucity of clearly stated expectations for evaluating proposed migrations using archaeological materials and other lines of evidence (Adams et al. 1978, McMillan 2003, Hughes 1992, Sutton 1991, Rouse 1986).

The strongest pairing of linguistic and archaeological data to evaluate past population movements comes when both lines of evidence are evaluated separately, on their own merits, and are then considered together (Blench 2006, Ehret 2012). For linguistics, such an evaluation might involve generating independent date ranges through the use of modern glottochronological-like approaches such as ASJP Chronology (Holman et al. 2011, chapters 1 and 9 [9.2]). For archaeology, it can entail predicting how language changes *might* look in the archaeological record and then assessing whether material evidence for them is observed.

Our study provides an opportunity to bring linguistic and archaeological studies together and evaluate these data through a critical lens. Linguistic evidence (9.2) suggests the movement of Proto-Utian (or Utian) speakers out of the Proto-Aleut-Utian homeland in coastal southwest Alaska (9.4.1) occurred sometime during the Middle Holocene. Loanwords indicate that Utians interacted with speakers of Wakashan, Salish, and Chimakuan languages during prolonged stopovers in coastal British Columbia and coastal northwest Washington State (chapter 6). Linguistic evidence further tells us that Utians eventually settled in central California, possibly centering on the greater San Francisco Bay Area, and that these people were the ancestors of later populations who spoke Ohlone (Costanoan) and Miwok languages (chapter 2).

In this chapter, we explore from an archaeological perspective the movement of Utian people from a proposed homeland in Alaska down the Pacific coast and eventually to the San Francisco Bay Area sometime in the Middle Holocene. We begin by scrutinizing the quality of the archaeological data on which our analysis is based, bearing in mind that in most instances the archaeological record lacks the temporal control and sample sizes for a robust comparison. We then lay out explicit archaeological criteria for evaluating potential contact between Utian peoples and those further south in British Columbia, Washington State, and California. We consult relevant literature on migrations to develop archaeological expectations for different scenarios for Utians moving down the Pacific coast. Based on our understanding of the regional archaeological record, we outline the kinds of data that may indicate contact between Utians and local coastal people. Our analysis thus provides both a methodological and a cultural-historical contribution.

Linguistic modeling of the Utian movement grounds our archaeological exploration both temporally and spatially. Temporally, the timeframe suggested by the ASJP model for the break up of PAU and its offspring (9.2, table 9.1) indicates that any archaeological indicators of the migration southward must date to the Middle Holocene. Also, based on the linguistic model, we anchor our exploration spatially on three areas that encompass the southward journey of the Proto-Utian-speaking people: the Greater Kodiak Island area of southwest Alaska—where the journey began; the Northwest Coast—where stops and cultural exchanges took place *en route*; and central California—where the voyagers eventually settled.

Within these three regions, our focus is on areas with robust and long-term archaeological records. In Alaska, we focus on the greater Kodiak region and, more specifically, the substantive research on Ocean Bay archaeology. In California, we focus on the greater San Francisco Bay Area with its well-documented shell-mound archaeology. In the Northwest Coast region, we draw on a sparse archaeological record associated with the Haisla and Heiltsuk territories of central British Columbia, and the Quileute and Upper Chehalis territories in Washington State. For each of these regions, we use our understanding of their distinct ecological, geomorphological, and social contexts during the Middle Holocene to evaluate the archaeological record for evidence of Utian migration and settlement.

8.2 ARCHAEOLOGICAL MODELS FOR POPULATION MOVEMENTS AND INTERACTIONS

Building on preceding literature, we identify three models of population movement and interaction that are relevant for examining the proposed Utian movement southward along the Pacific coast: population replacement, immigration, and population integration. Each model is associated with archaeological correlates that can be used to assess the possible presence of Proto-Utian speakers on the Northwest Coast and in the SFBA (see online appendix E, table E.1). A fourth possibility discussed in the literature, population movement to an unpopulated area (e.g., the Polynesian case in Rouse 1986:19–42), does not apply to the Northwest Coast or Central California since populations resided there prior to the Middle Holocene.

8.2.1 Population-Replacement Model

This model involves a sizable population moving into an area, bringing its own cultural practices, languages, and worldviews. In this scenario, the new population forcibly replaces the extant or resident population by displacing or dispersing them (Rouse 1986:177). Three archaeological expectations for this model are as follows. First, evidence for the material culture and cultural practices of the encroaching people should be apparent and should be distinct from that of the resident population. Earlier studies referred to this phenomenon as a "site-unit" intrusion (Willey et al. 1956:7–8, Rouse 1958:63–64) involving a complex of cultural traits (language, tools, architecture, burial modes, etc.) brought from the homeland to the newly occupied area (see also Rouse 1986, Heggarty 2014:618, Sutton 1991). Second, as recognized many years ago by Willey et al. (1956:25) and Rouse (1986:12), for the model to be supported, the intrusion of new cultural traits should replace previous archaeological complexes in the local region. Third, the arrival of the newcomers may be temporally correlated with evidence for warfare/violence associated with the removal or dispersal of the original resident populations.

8.2.2 Immigration Model

Rouse (1986:175–180) defines this process as when newcomers settle alongside resident communities, exemplified by his (p.177) discussion of the Island Carib movement into the Windward Islands in the Caribbean. The new settlers may retain their own language, cultural practices, technology, and architecture while maintaining a distinct colony or residential area (see also Spence 2005, Stein 2005). As with the replacement model, we expect this movement to be associated with the introduction of a complex of cultural traits from the immigrants' homeland. However, unlike the population-replacement scenario, the spatial distribution of these new materials should be restricted to specific spaces where the immigrants established their residences separate from the local populace. The resident population's material culture would continue relatively unchanged outside the immigrant residential areas. Such was the scenario in northern California with the movement of Native Alaskan workers to Colony Ross by the Russian-American Company to participate in the sea-otter fur trade in the early 1800s (Lightfoot et al. 1998), where the material culture of the Native Alaskans remained distinct from the Russians and Kashaya Pomo and Coast Miwok residents.

8.2.3 Population Integration Model

This model of population movement/interaction is not commonly articulated in earlier archaeological research but is gaining traction in more recent writings (Blench 2006, Heggarty 2014:616–618, Mallory 2021:286). Population integration takes place when a small number of travelers move into a new area and create relationships with the resident population that may result in the transmission of language, social and material culture, and technology. Social interactions between the travelers and the resident population may be created via marriage ties, exchanges, technological transfer, and religious practices. The newcomers may offer new kinds of subsistence practices, social organizations, ceremonies, and songs that resonate with and even enhance the prestige of some members of extant communities. Through these social networks, a convergence of languages and cultural practices takes place between newcomers and residents.

The archaeological expectations for population integration are as follows. First, while we expect to find evidence of new cultural traits arriving from the homeland of the travelers, we also expect these to be integrated into the material culture of the resident population. Thus, a continuation of the resident population's material culture would be enhanced with new material traits derived from a foreign homeland. The amount of material evidence associated with newcomers depends on a range of factors, including the length of interaction, the size of the newcomer population relative to the existing population, and the utility and distinctiveness of the new technologies. An inverse relationship between the relative size of the existing, resident population and the visibility of the material culture of the newcomers is expected.

8.3 PREVIOUS EXPLORATIONS OF THE NORTHWEST COAST AND CENTRAL CALIFORNIA

8.3.1 The Northwest Coast

For decades, Northwest Coast archaeologists have turned to population movement to explain culture change as well as the current distribution of ethno-linguistic groupings. These explanatory models were common from the late nineteenth to the mid-twentieth century, parallel to contemporaneous archaeology elsewhere in North America (McMillan 2003). Of special interest here, several of these early models attributed changes in artifact patterning through time to the in-migration of Alaskan peoples. By the late 1960s, however, migration theories fell out of favor in the Northwest Coast archaeological literature, again paralleling extra-regional patterns in archaeological thought. Locally, this shift in thinking was tied to the development of detailed local chronologies, as well as to an understanding that cranial deformation was the result of cultural practice rather than reflecting a different population (Matson and Coupland 1995, McMillan 2003).

More recent discussions of migrations and the archaeological record of the Northwest Coast are regionally focused and, importantly, rely on multiple lines of evidence, including ethnographic, ethnohistoric, linguistic, and archaeological data (e.g., McMillan 1998, 2003, Martindale and Marsden 2003). Archaeologically, they employ large and diverse datasets including settlement patterns, multiple artifact and feature types, faunal analyses, and several kinds of sites. Local knowledge, as recorded in ethnographic and oral historical documents regarding the identity of the newcomers, is an especially strong line of inference (see also Magne and Matson 2010).

Two examples of migration models in the Northwest Coast literature not only exemplify this diverse approach but also provide important modifications to the general models outlined above. Martindale and Marsden (2003) couple detailed and extensive ethnohistoric information with archaeological settlement patterns and artifact types to demonstrate that, between 3500 and 1500 BP, waves of inland Tsimshian (Wakashan) groups moved toward the settlement area of the Coastal Tsimshian. Such migrations occurred initially in small groups and are reflected in the introduction of non-local artifact styles and raw materials. No evidence of conflict is observed until later in this period, when movements southward by the Tlingit added to an already pressurized social-ecological system. This conflict, occurring over many centuries, eventually culminated in temporary abandonment, followed by social reconfigurations and changes in settlement patterns.

Similarly, McMillan (1998, 2003, see also Mitchell [1988, 1990]) finds support in the archaeological record for the linguistic model of the migration of the Wakashan-speaking people into the west coast of Vancouver Island. Specifically, he cites evidence in both the oral traditions of population movement and in a seemingly complete shift in artifact styles in some sites from those more typical of the Coast Salish region to those that are typical of later sites on the west coast of Vancouver Island. This transition in artifact styles is not accompanied by archaeological evidence of conflict, although this may be a sampling issue since population movements later in time are associated with conflict. McMillan (1998, 2003) suggests that the oral traditions and linguistic evidence of this proposed earlier migration are consistent with cultural integration. He also notes that a considerably larger sample size from multiple archaeological sites (including defensive sites) is needed to further scrutinize the hypothesis of population movement and its relationship to the linguistic models.

8.3.2 Central California

As with the Northwest Coast, migration as a causal mechanism in culture change is well integrated into California archaeological thought. Moratto (1984) presents a masterful synthesis of population movements in California from 12000 BP to 250 BP in his chapter "Linguistic Prehistory." He proposes distinct waves of people, who spoke different languages, expanding into new areas across California where they replaced prior residents, often forcing them to the margins of their previous homelands. Much of this synthesis is built upon linguistic reconstructions that are correlated with temporal changes in the archaeological record. In some cases, multiple lines of evidence are used to support the population-replacement model, as exemplified by the so-called Numic expansion in southern California and the Great Basin. Here Sutton (1991, 2014), Bettinger and Baumhoff (1982), and others employ archaeological, ethnohistorical, ethnographic, and tribal oral traditions to demonstrate the forceful expansion of Numic-speaking people into new resource patches.

More pertinent to the focus of this book, linguists and archaeologists have relied on each other's data to interpret the movement of Utians into central California. Citing linguistic data, scholars have proposed conflicting routes for this migration, including travel from the Plateau or Great Basin down the Sacramento Valley and into the SFBA (Breschini 1983, Golla 2007:75–76, Moratto 1984:552–557, Milliken et al. 2007:118, Whistler 1977:169–170), and possible maritime routes for Proto-Utian speakers that may have initially blazed into the SFBA and then spread eastward and northward, as noted by Bennyhoff (1994:65–66) and Fredrickson (see Milliken et al. 2007:118). Despite these different proposed routes, all agree that the first archaeological manifestation of Utians in the SFBA is associated with early shell mounds, specifically the so-called Berkeley Pattern, based on the initial study of the lower component of the West Berkeley Shell Mound (CA-ALA-307) (Bennyhoff 1994:66, Golla 2007:76, Moratto 1984:279, 554, Milliken et al. 2007:118, Whistler 1977:169).

Scholars have proposed the appearance of the shell-mound complexes to be associated with the first wave of Penutian speakers (speakers of Proto-Utian) who expanded into the SFBA about 5000–4000 BP (Golla 2007, Whistler 1977:169), supposedly replacing earlier Hokan or Yukian speakers. Moratto (1984:553) suggests that the "emergence of the Berkeley Pattern represents a fusion of older Hokan (mostly Esselen?) and intrusive Utian cultural elements in the Bay area." However, despite these assertions by linguists of the association of the *c.* 5000-year-old shell mounds with Utians, this has never been independently evaluated from an archaeological perspective. We conduct such an evaluation in this chapter.

8.4 ARCHAEOLOGICAL CONTEXT

8.4.1 The Greater Kodiak Region of Alaska: Ocean Bay Sites

Archaeologists use the name "Ocean Bay" to refer to a series of archaeological sites in the Kodiak Archipelago, on the Alaska Peninsula, and into Kachemak Bay, the area we propose as the homeland of Proto-Aleut-Utian people (9.4.1). The sites date between 7500 and 3500 BP and are thought to represent the earliest inhabitants of the region (Clark 1979, Fitzhugh 2003a and b, Kopperl 2003, Saltonstall et al. 2021, Steffian et al. 2006). This initial settlement follows what may be a time of low population numbers across much of the northern Pacific (Brown and Ames 2021). Our interest is in the oldest part of the Ocean Bay archaeological sequence (i.e., pre-5000 BP), since the Utians probably left their southwest Alaska homeland (9.2) and traveled southward sometime during this period.

While all Ocean Bay sites share similar characteristics, the bulk of our understanding comes from a few sites where extensive excavations have been conducted. Especially important is the Rice Ridge site in the Kodiak Archipelago, with its stratified, well-preserved, and well-dated archaeological deposits (Kopperl 2003, 2012, Hausler-Knecht 1993, Steffian et al. 2002). Stratified, deep deposits are important because they allow for more precise temporal control of changes in the material record. Such control is especially important in the case of the Ocean Bay sites since the archaeological tradition encompasses much time and cultural change that are beyond the temporal focus of this investigation. The stratified deposits and extensive dating allow us to focus on the earliest time frame (pre-5000 BP). The deeply stratified conditions of the Rice Ridge site also created soil conditions allowing animal bones to be preserved. Thus, while the other sites have almost no zooarchaeological record, the Rice Ridge site provides information on how animals were used for food and for raw materials (although a large portion of the bone-tool assemblage remains to be analyzed [Moss et al. 2021]). Moreover, since the site was excavated using a quarter-inch rather than an eighth-inch mesh screen, our understanding is biased toward larger faunal remains and against smaller bones such as those from herring.

The archaeological record tells us that the earliest residents of the archipelago were marine-focused people who were highly skilled sea-mammal and sea-bird hunters, fishers, and gatherers of clams and a range of plants. They lived in small settlements, sometimes year-round, in skin-covered tents, often with red ochre covering their living floors (Moss et al. 2021, Steffian et al. 2002, Hausler-Knecht 1993, Fitzhugh 2003a and 2003b:44, Clark 2001:154, Saltonstall et al. 2021). During the earliest times, the material record was focused on blade (large and small) technology and minimal ground stone, made of both off-island and local materials. Later in time, the prevalence of blades and non-local materials decreases, with an increasing reliance on bifacial tools from local sources (Fitzhugh 2004, Steffian et al. 2002). By *c.* 6000 BP, we see an expanded use of ground slate technology (Hausler-Knecht 1993, Saltonstall et al. 2021).

Despite its richness, the extant Ocean Bay archaeological record is limited from our perspective since it does not encompass the full temporal scope of this inquiry, that is, the possible first 1,500 years (9000–7500 BP) of Proto-Aleut-Utian life may not be recorded in the sites examined. Thus, if people left their homeland any time before 7500 BP, our investigations would include no material evidence of their movements.

8.4.2 The Northwest Coast

Middle Holocene archaeology on the Northwest Coast has received scant attention and is not well-understood (Moss et al. 2007). Archaeologists have instead focused on both ends of the culture-historical continuum: on the earliest sites in the region, dating back some 14,000 years (McLaren et al. 2015, 2020), and the Late Holocene sites often characterized by well-established and extensive village sites embedded within complex social-ecological systems (Lepofsky et al. 2024). The Middle Holocene is a culturally significant time, when groups in many parts of the coast increasingly settled into and integrated with their surrounding land- and seascapes. All Northwest Coast peoples have oral traditions that outline these deep-time connections to their current territories, including their origins through ancient migrations and transformations from spirit beings.

8.4.2.1 Central Northwest Coast, Heiltsuk, and Haisla Territories

We have only spotty information on the Middle Holocene archaeological record of Heiltsuk and Haisla traditional territories and how people lived in these outer coastal, inlet, and riverine environments. To our knowledge, no full-scale excavation of sites in Haisla territory has been carried out, and the archaeological record for this period is relatively unknown. In contrast, research has been extensive in Heiltsuk territory, mostly focused on regional site surveys, with limited below-ground explorations. These surveys have produced the oldest and longest-occupied settlements on the Northwest Coast (McLaren et al. 2019). The 14,000-year-old settlement on Triquet Island (McLaren et al. 2015, 2020) produced extensive lithic cores and debitage (waste flakes), and faunal and botanical remains (D. McLaren, pers. com. 2023). From its earliest levels, the site was the home of generations of people who were well adapted to their marine and terrestrial landscapes, using a range of resources from diverse ecosystems. The extensive surveys, while not providing detailed artifactual evidence, overwhelmingly demonstrate that, when they traveled southward sometime in the Middle Holocene, Utians were met *en route* with well-established cultural groups, with complex social and ecological systems that were inextricably linked to their land- and seascapes (McLaren et al. 2015).

For artifactual evidence, the famous archaeological site of Namu in Heiltsuk territory provides rich and diverse assemblages. Large-scale excavations through the deep deposits were conducted there in broad temporal and stratigraphic layers in 1969, 1970, 1977, 1978, and 1994 (Carlson 2014). The site dates to some 11,000 years ago, with the cultural sequence described in large temporal swaths. Of interest to us here are period I (11000–7000 BP) and periods II and III (7000–5000 BP). The later excavations at Namu were conducted using a fine (one-eighth-inch mesh screen), producing a rich faunal record (e.g., Cannon 1991). However, faunal remains are preserved only in period II and onward.

8.4.2.2 Southern Northwest Coast, Quileute, and Upper Chehalis Territories

Only a very sparse Middle Holocene archaeological record is found for the Quileute on the Olympic Peninsula and the Upper Chehalis along the inland rivers. Little is known about the pre-3000 BP archaeological record of western Washington State, in general due to sea-level changes, development, lack of chronological controls (Schalk 1988), and, in earlier explorations, the use of coarse- rather than fine-mesh screens (e.g., Greengo and Houston 1964). According to archaeologist Robert Kopperl (pers. com., 2023), most sites in the river basins older than the last few thousand years are unstratified and have been dated based on artifact styles that may not be temporally distinct, rather than with radiocarbon analyses (e.g., Welch 1983). He notes, however, that better survey methods and more focused sampling efforts are beginning to produce Middle Holocene radiocarbon-dated sites with well-preserved remains. Sites along the river mouths are relatively recent, since older sites are buried due to local subsidence and delta progradation. Placing sites in the older time periods has often been based on the recovery of sometimes serrated, large bifaces, termed "Olcott Points." Consequently, we have little direct archaeological evidence from this region that allows us to examine the material correlates of cultural exchanges with Utian travelers.

8.4.3 The San Francisco Bay Area

The dynamism of the SFBA landscape in the Middle Holocene (see chapter 7) not only affected human settlement but also influences our understanding of these past peoples as seen through the archaeological record. Few sites dating to the Early and Middle Holocene have been identified in the region, and little is known about the early people outside of isolated finds and occasional human burials (Byrd et al. 2017:7–1 to 7–9, figure 33). This paucity of knowledge is almost certainly related to the flooding of the greater San

Francisco Bay from 10000–6000 BP and the subsequent burial of sites situated along the former riverine system under many meters of alluvium (Byrd et al. 2017:7–9, Meyer and Kaijankoski 2017, Rosenthal and Meyer 2004b:1–9). Such was certainly the case in nearby valleys, where cycles of landform stability were interrupted by episodes of widespread alluvial deposition associated with Early and Middle Holocene climate change (fluctuations in precipitation and changes in vegetation cover) that buried ancient sites (Meyer and Rosenthal 1997:II.1–II.11, Rosenthal and Meyer 2004b:11–35). The combination of the emergence of the broader San Francisco Bay estuary and the periodic burial of early inland sites under substantial alluvial deposits limits our ability to detect sites in the SFBA older than 6000 BP, that is, as much as 2,000 years after the start of the focal period of this inquiry. This point is exemplified by the recent discovery of a deeply buried site in the historic Mission Bay in San Francisco, where geoarchaeological coring as part of proposed development unearthed a 7900–7300 BP shell midden under 12 meters of bay mud. While the coring recovered evidence for shellfish (oysters) and archaeobotanical remains (some nuts and seeds), little is yet known about the artifactual assemblage, given the difficulties of investigating a site buried this deep (Redmond 2023).

To expand our temporal gaze, we turn to two of the best-documented buried inland sites in the SFBA that date prior to the establishment of the shell mounds. These Early Holocene and early Middle Holocene sites probably date prior to the Utian migration to the area and may provide material evidence of its original "resident" population. These sites were unearthed during recent archaeological investigations involving major earth moving via cultural resource management projects. CA-CCO-696, the oldest site, is situated in the upper Kellogg Creek valley in southeastern Contra Costa County. It was discovered during mitigation work undertaken for the construction of the Los Vaqueros Reservoir. Archaeologists revealed the oldest component of the site associated with the Kellogg Paleosol during their "Deep Exposure Excavation." The archaeological investigation of the Kellogg deposit uncovered one burial, no cultural features, and a modest-sized assemblage of artifacts, zooarchaeological specimens, and archaeobotanical remains that date between 9870 and 7400 cal. BP (Meyer and Rosenthal 1997) (online appendix E, section E.a).

The second site, CA-CCO-18/548, is in northeastern Contra Costa County on the edge of Marsh Creek, which flows from Briones Valley. Situated on both public and private land, archaeologists conducted excavations to stabilize the creek bank on State Park property and to mitigate the impact of a major residential development on private property (Rosenthal 2010, Wiberg 2010). We focus on findings from the excavation of the lowest and earliest deposit of the site, identified as Stratum II, dating from 7060–5025 cal. BP.

The investigation of Stratum II unearthed three burials, four cultural features (a baked clay hearth, two fire-cracked rock features, and a cluster of water-worn cobbles), and a significant assemblage of artifacts, zooarchaeological remains, and archaeobotanical remains.

While the archaeological assemblages from these two Early/Middle Holocene sites vary somewhat, their many similarities suggest they are part of a broader cultural tradition (see online appendix E, table E.2). Both assemblages include similar lithic tools (e.g., large projectile points, battered cobbles, and milling slabs and hand stones), a wide array of plants (e.g., acorns, grass seeds), large and small game, and fish (at CA-CCO-18/548) that indicate the harvesting of resources from nearby riverine grassland-savanna, and oak woodland environments. Both appear to be seasonal camps established along interior drainages by local groups, probably in the late spring, summer, and fall.

After 6000 BP, when the modern San Francisco Bay landscape began to take shape, the people of the SFBA initiated construction of imposing (and archaeologically visible) shell mounds (online appendix E, section E.b). These sizable mounds reflect multi-generational, deep connections to the SFBA landscape on the part of Native Peoples. The earliest deposits in the mounds, while not well-understood archaeologically, have the potential to provide insights into how people settled into the newly created bay environments. The SFBA mounds comprised many cubic meters of sediments, marine shells, rock, and ash constructed over many centuries by multiple generations of people and are part of a broader manifestation of elaborate, shell-rich anthropogenic landscapes created by people across the globe in coastal and riverine environments (e.g., Fish and Fish 2010, Lightfoot 1997, Lightfoot and Luby 2002, McNiven 2013, Sassaman 2004).

The primary field studies of the mounds took place in the early 1900s, before many were destroyed by urban development along the bay shore. Investigations were made using coarse-grained methods that involved shovel broadcasting with little or no use of screens, the rapid excavation of extensive areas of some sites, and the collection of large formal tools and large-bodied faunal remains, most of which were provenienced to specific stratigraphic locations. These early explorations produced a wealth of material concerning the constituents of the shell mounds, including human burials, extensive faunal assemblages of fish, marine and terrestrial mammals, and birds, a diverse range of artifacts, and some architectural features. Later analyses of sediment samples collected by early excavators have revealed a diverse range of small-bodied faunal remains, micro-artifacts and debitage, and plant remains (online appendix E, section E.b).

The archaeological assemblage of the oldest known shell mound, West Berkeley (CA-ALA-307), with the lower mound component dating to

6000–3100 cal. BP, comprised the so-called Lower Berkeley Pattern associated by some scholars with the movement of Utian people into the area (e.g., Bennyhoff 1994, Golla 2007, Moratto 1984, Whistler 1977). While important and valuable in its richness, the lack of temporal specificity in the basal component (a 3,000-year spread), and the fact that it only overlaps with our focal time period for 1,000 years, unfortunately limits the value of this data set in our exploration of evidence for Proto-Utian speakers in the ancient SFBA.

8.5 METHODS: ARCHAEOLOGICAL EVIDENCE FOR UTIAN MIGRATION

Our exploration focuses on Utian migration to those regions with the strongest linguistic evidence for visitation (central and southern Northwest Coast) and eventual settlement (central California), and where the most complete archaeological record is found. Building on our summaries of the regional archaeological records provided above, we lay out criteria for evaluating similarities and differences among the assemblages in Alaska, central and southern Northwest Coast, and the SFBA. We are mindful that all the records in the four regions are limited by varying issues of sample size, recovery methods, and temporal control.

We use two approaches in our search for material evidence of Utian southward migration. First, we examine whether broad evidence for any of the migration models presented earlier (and in online appendix E, table E.1) is apparent. Second, we examine in detail whether any specific artifact types of more northerly sites are found in more southerly sites and in the SFBA, the new Utian home.

8.5.1 Archaeological Evidence for Migration Models

Although the SFBA data are limited temporally, we reason that this region, where Utians settled and where a somewhat large artifactual inventory is found, is the most likely to show evidence of the newcomers. Since the mounds themselves date only to 6000 BP and later, we expand our temporal range by including the two earlier inland sites (CA-CCO-696 and CA-CCO-18/548, *c.* 9900–6000 BP) that may offer material evidence for the original residents of the greater SFBA. If the Utian migration made a substantial impact on material culture, we should see it through the comparison of these three assemblages. More specifically, if Utians were responsible for building the mounds, as suggested in earlier studies, then the material culture of the earliest levels of the mounds should look distinct from those of the earlier inland sites. On the other hand, similarities in the assemblages of the two inland sites and the earliest levels of the Bay Area mounds (9900–3000 BP) might indicate the latter were built (at least in part) by people already in the region.

While our comparison encompasses the period of interest, several differences in the archaeological context of the two sets of assemblages limit the inferences that can be made. First, the ecological contexts of the two sets of sites are very different, which would have offered distinct opportunities and restrictions to the sites' inhabitants. The West Berkeley mound was constructed on the shores of San Francisco Bay, while the two interior sites were situated in grassland-savanna and oak woodlands along drainages that flowed into the bay/delta. Relatedly, the interior sites appear to be seasonal residential bases and the West Berkeley mound, at least in later times, is thought to be a long-term mounded village associated with diverse ceremonial, residential, and subsistence activities (Lightfoot et al. 2022). Thus, any differences observed in the material records could simply be related to people making different choices about how best to live in those places, rather than reflecting differences in origin.

Furthermore, the size and nature of the sample of remains recovered from the two sets of sites are not fully comparable. Specifically, the sample of material from the deeply buried interior sites is smaller and less diverse than that from the mound (online appendix E, sections E.b and E.c). The number of artifacts (not including lithic debitage) for CA-CA-CCO-696 is forty (Meyer and Rosenthal 1997: table III.5) and for CA-CCO-18/548 is sixty-three (Rosenthal 2010: table 75), while the total artifact assemblage for West Berkeley is calculated at over 4000 (Wallace and Lathrop 1975:8). Furthermore, preservation of faunal remains is likely better in the shell-rich (high pH) deposits of the bay-shore shell mound, although a wide range of faunal remains was recovered from all sites. Finally, since the basal deposits in the mound span 3,000 years (the latter of which post-date our time of interest), differences in the two sets of assemblages may be due to the conflation of later material evidence not relevant to our Utian migration proposal.

8.5.2 Artifact Types of the North Found in the South

To explore potential archaeological evidence of the Utian southward movement in greater detail, we scrutinize the archaeological records of the Ocean Bay Tradition of Kodiak Island, Alaska, as well as those from the central and south Northwest Coast, and from the SFBA for artifacts or other material remains that may have been brought from Alaska or the Northwest Coast to the SFBA. We limit our investigation to the Middle Holocene.

We use several criteria to help us more rigorously examine the archaeological record. For instance, in many cases, we avoid drawing conclusions based on similarities and differences in plant (paleoethnobotanical) and animal (zooarchaeological) remains across the regions, since huge discrepancies in sampling effort and in preservation across the regions prevent meaningful comparisons. For instance, no paleoethnobotanical analyses have been conducted at the Alaska and Northwest Coast sites. Similarly, the sites were not all excavated with attention to stratigraphic details reflecting temporal differences, nor was a fine-mesh screen consistently used to retrieve small faunal remains. Moreover, the absence of evidence of plant use or the extensive harvesting of herring is almost certainly one of sampling bias, not cultural difference. Differences in the preservation of faunal remains will greatly influence our understanding of human-animal interactions. Like the differences in the faunal assemblages at the non-mound and mound sites in the SFBA, the lowest levels of many sites further north have poor faunal preservation due to low pH (high acid) conditions. Early/Middle Holocene Ocean Bay deposits (Ocean Bay I) have poor bone preservation (Patrick Saltonstall, Alutiiq Museum, pers. com. 2023, Hausler-Knecht 1993), greatly influencing our ability to understand how people used animals as a source of food and resources (i.e., bones for tools). We assume, in general, that the differences observed in plant and animal remains across the three regions are largely the result of these differences in sampling and preservation, rather than cultural preferences for harvesting specific species.

Ecological differences in the three regions present different opportunities for people to interact with their biological worlds. As a result, different local technologies will have developed to harvest and process these plants and animals. For instance, since acorns are restricted ecologically to the southern Northwest Coast and coastal California, the fact that they are not found in the archaeological record of the northern Northwest Coast or Alaska is not culturally meaningful in this context. Similarities and differences in technology associated with resources that are widespread, such as large sea-mammal hunting, however, might be a meaningful pattern to trace if we were we able to control for sampling and preservation.

We are on much firmer ground when examining similarities and differences across the three regions in the forms of lithic (stone) artifacts, especially the common forms. The preservation of stone artifacts (unlike those of bone) is unaffected by most soil conditions, and the recovery of lithics is often targeted in archaeological excavations, regardless of excavation strategy. Thus, most excavations will produce an adequate sample of the common forms of lithic artifacts. Making cultural inferences from the presence or absence of a rare type of artifact is much trickier, since it is harder to evaluate whether the presence of a rare form is culturally meaningful or merely the result of sampling bias without a large sample of excavated material.

Of the lithic artifacts, we are most interested in those diagnostic of specific times and places. We seek to identify unique artifacts found in northern sites and whose spatial and temporal distribution is consistent with Utian southward travels. Consequently, we exclude from our analysis common forms of artifacts that were used coast-wide (and sometimes worldwide) because they are a useful tool form, as is the case with atlatl throwing boards. While archaeological evidence exists of these wooden spear throwers in the early Ocean Bay sites in Alaska (Steffian and Saltonstall [2024]), as well as on the central BC coast (McLaren et al. 2015, 2019, Carlson 2014), our knowledge of their specific forms is insufficient to claim this temporal co-occurrence to be anything more than a local expression of a widely used and valued tool type.

Our analysis also excludes artifacts displaying a region-wide, time-transgressive distribution from northern to southern areas since their distribution can be explained parsimoniously by the slow, down-the-line exchange of ideas and material culture by coastal peoples. Such is the case with the generalized use of ground slate technology found in Ocean Bay sites from *c* 6700 BP and then in progressively younger contexts southward through the Northwest Coast (Ames et al. 2010, Dinwiddie 2014), but are not identified, with a few possible exceptions, in SFBA sites. Microblades are found throughout the archaeological record of the north coast until about 5000 BP and to at least 1500 BP further south along the coast (McLaren 2008, 2017), but are not commonly found in SFBA sites. Thus, our best candidates for material culture associated with Utian southward movement are artifacts that are stylistically unique, come from well-dated archaeological contexts, and have discontinuous distributions consistent with Proto-Aleut-Utian/Proto-Utian loanwords.

Understanding the context in which artifacts were used and deposited is also key to this analysis. For instance, red ochre is used throughout the region (and the world) for a variety of purposes (MacDonald 2008), and in the Ocean Bay site of Rice Ridge, it was uniquely spread on house floors (Fitzhugh 2004). Ochre has also been recovered throughout the Northwest Coast sites and the SFBA, but the context of recovery is different, e.g., burials and isolated nodules in the midden layers. Thus, inferences about culture contact between Utians and other groups on the coast based on the generalized use of ochre would be inappropriate.

Differences in archaeological house forms are excellent indicators of past identities (e.g., Lepofsky et al. 2009), and were one of the criteria used to identify the migration of the interior peoples to Tsimshian territory in the example provided earlier (Martindale and Marsden 2003). Except for the unique mounds in the SFBA, however, we have no archaeological information on Middle Holocene house forms for the regions in our study area.

8.6 RESULTS

8.6.1 Archaeological Evidence for Migration Models

Despite differences in chronology, archaeological contexts, sample sizes, and geography, our examination of the archaeological records of the two inland sites (CA-CCO-696 and CA-CCO-18/548) and the West Berkeley mound indicates enough basic commonalities to suggest at least some shared cultural practices. Specifically, we find that all three assemblages contain stemmed projectile points, battered cobbles and cobble tools, bifaces, cores, modified flake tools, and some debitage (recovered from recent flotation analysis of West Berkeley, online appendix E, table E.2). Furthermore, even though the sites are situated in different ecosystems, similar species of both large and small terrestrial mammals are represented. However, consistent with its location in a coastal setting, the West Berkeley's faunal assemblage also contained a diverse range of marine mammals, marine fishes, and water birds not found in the interior sites (online appendix E, section E.b). The flotation of sediment samples from the three sites demonstrated acorns to be a consistent food source, along with other nuts and seeds.

Differences in the assemblages between the inland and coastal sites are explained most parsimoniously by their different environmental contexts, site function, and the coastal mound conflation of 3,000 years. For instance, the ground-stone assemblage of the interior sites is limited and includes only milling slabs and hand stones, whereas the coastal mound assemblage contains mortars and pestles as well as charmstones and other categories of ground-stone artifacts. Differences in some ground-stone tools may reflect different technologies employed in the processing of plant foods in the interior and coastal environments, even though acorns were valued in both contexts. They may also reflect the later age of the West Berkeley assemblage and innovations taking place across central California in milling tools (e.g., the adoption of mortars and pestles) after 5500 BP (Byrd et al. 2017: 10–6).

Differences in bone assemblages in the inland and coastal sites may reflect a more diverse range of baskets, textiles, fishing nets, and other cultural items produced at West Berkeley. Only a small sample of worked bone artifacts was recovered from the interior sites, while the West Berkeley deposits yielded a diverse range of bone tools, including awls and needles (online appendix E, section E.b). Given that some fauna are preserved in the interior sites, this difference is unlikely to be accounted for by preservation or sampling issues alone, but probably reflects the fact that the inland sites are camps and the mound sites are more permanent settlements.

Taken together, similarities and differences in the assemblages of the interior and mound sites suggest both continuity of cultural practices and the adoption of new cultural practices over time. The evidence for continuity and the absence of any significant evidence of conflict do not support the population-replacement model that predicts a significant number of Utian people arrived in the SFBA and completely replaced the resident population (see online appendix E, table E.1, online appendix E, section E.c). Moreover, we find little evidence in the West Berkeley mound for the immigration model, where new cultural practices were established separate from those of the resident population. Rather, the continuity in chipped stone tools and cultural salience of acorns and small seeds suggests the people who created the West Berkeley mound built upon a cultural repertoire already used by people residing in the region. They enhanced these core cultural practices with innovative material culture (an extensive bone-tool and ground-stone technology) and subsistence pursuits (a focus on marine fauna) that is present in the lower component of West Berkeley. We now examine whether these new cultural practices resulted from the arrival of Utian peoples, who integrated their material culture with that of the resident population.

8.6.2 Artifactual Evidence of Utian Migration

To further investigate the archaeological correlates of Utian southward migration, we examine the artifact assemblages from the study areas, beginning with outlining the temporally and stylistically diagnostic artifacts from the earliest Ocean Bay deposits in Alaska (see online appendix E, table E.3, column 1). We scrutinize the Middle Holocene archaeological records of central and southern Northwest Coast and the SFBA to see if any of these distinctive artifacts are found further south (online appendix E, table E.3). For all areas, given the limitations of the archaeological record (see above), the absence of a certain artifact type may not reflect its true absence from the regional assemblage. This is especially true for the southern Northwest Coast, where the early sites are largely limited to poorly dated surface finds. Sampling and preservation biases and temporal inconsistencies, sometimes involving tens of hundreds of years, however, occur throughout the geographic scope of this study.

Given these caveats, it is understandable why the assembled data (online appendix E, table E.3) do not include material evidence for contact between Utians and the ancestral Heiltsuk, Haisla, Quileute, or Upper Chehalis peoples. For similar reasons, the data include only limited and ambiguous support for the proposal that the SFBA mounds were constructed by Proto-Utian speakers.

Although the SFBA is the most likely area for finding material evidence of Utians since this is where they eventually settled, very few of the diagnostic Ocean Bay lithic artifact types have been detected in SFBA sites. The shell mounds lack the microblade and microcore technology found in the North Pacific and the ground slate tradition that became increasingly important in later Ocean Bay sites. The few diagnostic lithic types found in SFBA are a few bar- or pencil-shaped ground stone implements fashioned from green and brown serpentine recovered in the lower component of West Berkeley, and possibly some of the large bifaces.

Some similarities exist between the worked bone artifacts found at West Berkeley and Ocean Bay sites, especially with the sophisticated development of various kinds of awls, needles, and other tools probably used for making baskets, textiles, and possibly nets. A significant difference in the worked bone assemblage is the paucity of diagnostic harpoon points in the shell mounds of SFBA (see online appendix F, table E.3).

8.7 DISCUSSION

8.7.1 Evaluation of the Three Migration Models

Archaeological investigation suggests little direct evidence for support of the population replacement or the immigration models for the movement of Utian people into SFBA in the Middle Holocene. Instead, findings indicate that some degree of continuity took place between the resident populations of SFBA and those who built the lower mound at West Berkeley from 6000–3100 BP. Furthermore, little evidence of interpersonal violence or warfare is associated with West Berkeley or other early shell-mound sites that might be associated with population replacement (online appendix E, section E.c).

Of the three models, the population-integration model is most in accord with reviewed data. Such integration may have resulted from a shared cultural focus on the marine-scape that would have encouraged the transmission of knowledge and cultural practices, as well as the development of a shared language, between newcomers and residents when they both colonized the newly emerging San Francisco Bay in the Middle Holocene. A shared maritime focus is demonstrated by evidence that Utians and the people who built the West Berkeley mound harvested many of the same water birds, such as geese, ducks, cormorants, and murres. Maritime fishing was also important throughout the region and, while the species varied in the North and Central Pacific waters, all fished for salmon and herring.

The maritime people who constructed West Berkeley also hunted many of the same sea mammals as those found in Alaska and Northwest Coast waters, including sea otters, harbor seals, and sea lions (words for which are reconstructed for Proto-Aleut-Utian and Proto-Utian, see chapters 2 and 9 [9.4.1]). We note, however, that the distinctive sea-mammal hunting technology of the Native Alaskans, including harpoons and slate knives, did not seem to travel southward. Thus, while we do not find direct evidence that residents borrowed hunting technologies from Alaska, we imagine the exchange of ideas and posit that Utian voyagers adopted some hunting practices from residents and may have developed new ones in concert with them. Such cultural integration is consistent with Moratto's (1984:553) idea of the fusion of the resident Native California inhabitants of the SFBA with Utian newcomers. What differs from Moratto's and other previous interpretations is the infusion of people and new cultural practices derived from Alaska and the Northwest Coast. This finding suggests that Proto-Utian speakers in California may have ancestral roots that go back many thousands of years in both the North Pacific and Central California.

8.7.2 Factors Limiting Detection of Artifact Similarity Between North and South

Comparisons of the material records of Alaska, the Northwest Coast (BC and Washington State), and the SFBA have not produced conclusive archaeological evidence for the Utian migration sometime between 9000 and 5000 BP. This is not surprising if the migration took place during the very earliest years of the period, that is, at a time not consistent with existing archaeological records. While some tantalizing similarities in material culture among the four regions (e.g., ground rods and bars, large lanceolate bifaces) are found, far more differences among them exist. Furthermore, some similarities can be explained by coastally shared cultural traits reflecting people's adaptation to, learning from, and managing their surrounding marine and terrestrial landscape.

Archaeological explorations of migrations can also be hampered when trying to compare different archaeological contexts, especially those from diverse temporal and spatial settings, and inventories from different sites will be influenced by differential preservation, recovery methods, and sample sizes. In an ideal world, the evaluation of population migrations using archaeological data should embrace a large sample of diverse materials (i.e., micro-lithics, worked bone, antler, and shell artifacts, faunal specimens, and botanical remains) recovered with standardized methods. In our case, the disparate contexts of our samples make it impossible to fully compare across regions and, as a result, we focus only on a small subset of the artifacts recovered from Ocean Bay sites (online appendix E, table E.3).

In the case of the SFBA shell mounds, pioneering investigations in the early 1900s were undertaken using coarse-grained methodologies that exposed extensive areas of sites with no systematic use of screens. While a large sample of large formal tools and faunal specimens was recovered, many small-sized artifacts, small-bodied fauna, and botanical specimens were missed. The coarse-grained excavation methods may have precluded the recovery of microblades and microcores that are diagnostic artifacts of Ocean Bay sites. Later investigations of shell mounds using fine-grained methodologies (e.g., flotation and small mesh sizes) are now providing information on many of these missing classes of archaeological materials, but the volume of the deposits examined is minimal compared to the earlier excavations. These fine-grained, small-scale investigations may not yield representative samples of large formal artifacts, particularly those that are relatively rare and diagnostic, such as bone harpoons, ground slate bayonets, and ground slate rods. Thus, in the future, very different interpretations about population migrations may be generated based on the presence of diagnostic cultural materials depending largely on the degree of organic preservation, recovery methods involved, and volume excavated.

The paucity of diagnostic Utian materials in Northwest Coast and SFBA sites may be related to spatial limitations on items that could be carried onboard watercraft that traveled down the Pacific coast. Ancient watercraft would have had limited storage room for cultural materials, and thus, the travelers may have only brought with them a small subset of their material culture.

Such was the case with the Alutiiq (Yupik) people (mostly from Kodiak Island, Alaska) who traveled to Colony Ross (California) to work as sea-mammal hunters for the Russian-American Company in the early 1800s. Excavations of the Native Alaskan Village Site (NAVS), where the Alaskan sea-mammal hunters resided when not working (hunting) for the Russians, revealed evidence of their sophisticated marine hunting and fishing equipment, especially large and small darts, harpoon arrow points, harpoon shaft (socket) pieces, and composite fishhooks. Most surprisingly, however, little to no evidence was found of Alutiiq household equipment or furniture that was used in contemporaneous Alutiiq villages on Kodiak Island, including stone lamps, ground slate ulus, ground slate knives, adzes, bone spoons, Alutiiq pottery, and hollowed-out whale vertebrae platters (Lightfoot et al. 1998). Thus, many "diagnostic" Alutiiq cultural materials found in archaeological contexts in Alaska were not transported to/re-created in California, which may be related to the nature of the work (e.g., sea-mammal hunting) at Colony Ross, and that the majority of the workers were men (although some Alutiiq women also resided at NAVS), and what was allowed to be transported on Russian ships that carried the workers to the area. Furthermore, the common formation of interethnic households composed of Alutiiq men and California Indian women resulted in the adoption of local Indigenous domestic equipment and furniture that took the place of traditional Alutiiq household materials.

Utian travelers to California may have been constrained by the size of their watercraft, the size and composition of the traveling party, and the kinds of tasks that were considered essential in ocean voyaging (chapter 9, 9.5.4). A limited number of materials, such as harpoon points, ground slate tools, and so on, were probably conveyed from Alaska to the Northwest Coast and to California. Whether these cultural items were then refabricated at stops on the Northwest Coast or in California would have depended on the tasks undertaken, whether local sources of raw material for producing the tools were available, and the cultural repertoire of the resident populations with whom they interacted. The production of ground slate tools in the SFBA may have been limited, as local sources of high-grade slate applicable for tool making are not common, while the manufacture of bone harpoon points may have been constrained by the amount of sea-mammal hunting undertaken in bay waters and the methods of hunting employed by resident people (e.g., hunting with nets). As with the Colony Ross example, interactions with local Indigenous residents of SFBA and a growing familiarity with their cultural practices may have diminished the necessity of re-creating some diagnostic Ocean Bay objects in the new homeland.

Chronological control is a foundational challenge associated with the study of ancient migrations. In our study, well-dated sites of the crucial periods for tracking the Utian southward movement are few, due to the complex geomorphological histories of the study area as well as, in some cases, the lack of concerted effort to fully examine sites dating to this time period. Thus, in the greater Kodiak Island region, no detailed archaeological record is associated with the potentially earliest time of the migration (9000–7500 BP). Similarly, few well-understood sites in the northern Northwest Coast and SFBA and none in the southern Northwest Coast date to this time. Thus, 1,500 years of our migration story are missing from the archaeological record.

An avenue for future research is to broaden the archaeological investigation to include other potential places within the Proto-Aleut-Utian homeland in coastal south Alaska. While this chapter has focused on Ocean Bay archaeology of the greater Kodiak Island region, the archaeological record of other nearby areas could also be studied in depth. This point is exemplified by Corbett and Hanson's (2023) recent masterful synthesis of the rich history of the Unangax̂/Aleut people of the Aleutian Islands. They document an archaeological record in the eastern Aleutian Islands that extended back to at least 9000 BP, which could provide archaeological materials associated with the earliest time of the Utian migration (9000–7500 BP).

Similarly, a 1500-year disconnect exists between the earliest known record in the greater Kodiak Island region (i.e., *c.* 7500 BP) and the earliest well-documented archaeological layers in the SFBA (i.e., *c.* 6000 BP). The period between 6700 and 6000 BP is marked by the gradual decline of microblade artifacts among Ocean Bay communities and their experimentation with the making of ground slate rods and long, slender bayonets (Saltonstall et al. 2021:3.4, 4.3). If the Proto-Utian migration to California took place prior to 6000 BP, however, the lower component of West Berkeley might not contain these diagnostic artifacts but, rather, the deposits would represent a later occupation, probably several generations after the arrival of the Alaskan Natives when only a few vestiges of their material culture (ground-stone rods, large bifaces) might still be in use. Future cultural resource management (CRM) fieldwork along the shoreline and submerged landscape using deep excavations of buried sites may detect earlier archaeological contexts, providing the opportunity to examine this critical pre-6000 BP period and evaluate whether diagnostic Ocean Bay materials exist. The recent detection of the 7900–7300 BP shell midden site in historic Mission Bay in San Francisco, buried 12 meters below bay mud deposits, indicates the existence of these ancient sites (Redmond 2023).

8.8 CONCLUSION

Although our exploration does not produce unambiguous archaeological evidence for the Utian southward movement, it does provide a range of parameters to guide future studies. Such studies must be based on large, well-dated samples from a variety of archaeological sites in both the homeland and the places visited along the way. Robust archaeological interpretations of past interactions can then be made and compared to linguistic evidence. Such interpretations are made even more robust when other lines of evidence, such as ethnographic and oral historical evidence, are examined for independent evidence of migrations. When critically combined, such investigations have the potential to offer many insights into how people respond to novel ecological and social settings when moving through a complex world. Future studies using archaeological remains and based on the above-discussed parameters may provide new insights on the movement of Utian and other peoples along the Pacific coast.

Chapter 9

Culture History Reconstruction

9.1 INTRODUCTION

We follow the social and environmental pathways of Utians on their journey southward from their Alaska homeland at some time during the Middle Holocene. We explore their migration, especially through evidence preserved in their language and in those spoken by Native Peoples encountered along the way. We also gather archaeological, ecological, and genetic information attesting to the timing, nature, and route of their trek and eventual settlement in the San Francisco Bay Area of California. In this chapter, we bring all threads of the story together to expand and enrich our reconstruction of this epic journey. Relying on inferences based on both factual detail and, more generally, on knowledge of the region and its broader cultural history, we provide a narrative of an important time in prehistory featuring the lives of many peoples untold until now.

9.2 ASJP CHRONOLOGY

Important to our reconstructed story is a timeline to which all parts of the narrative are related. The timeline used here is informed by ASJP Chronology. ASJP is introduced in section 1.2.3. More information can be found at The ASJP Database (clld.org). The approach was developed for automated language classification using binary comparisons of languages in the ASJP Database (of which there are 5,590). For each binary language comparison (two compared languages), a lexical similarity index is generated. This index is a similarity percentage for the two compared languages ranging from 100 percent (lexically entirely the same) to lesser percentages including 0 percent (no lexical similarity at all). Calibrating percentages with known historic, epigraphic, and archaeological dates for language divergence yields a standard against which the date of divergence can be accurately estimated for any language pair. (Two language dialects are considered to have completely diverged from a common parent language when the two codes are recognized as distinct languages.) For more details, consult Holman et al. (2011), in which ASJP Chronology was originally described and applied.

Once a similarity index is determined for a language pair, a language divergence date is calculated by entering the value of the index into a formula described in Holman et al. (2011:844). For example, the value of the index for the comparison of Proto-Costanoan and Proto-Miwok is 8.29. (Proto-Costanoan and Proto-Miwok are the two and only direct daughter languages of Proto-Utian.) Using the formula, this value yields a date of 3663 BP, the estimated number of years before present that these two languages diverged from Proto-Utian, their common parent language.

Holman et al. (2011) observe discrepancies between estimated and calibration dates for ASJP that are, on average, 29 percent as large as the estimated dates themselves. We deal with these discrepancies by calculating for each anchor date (e.g., 3663 BP for Proto-Costanoan/Proto-Miwok) a range of dates within 29 percent of the estimated date. This range is calculated as follows, using as illustration statistics for the Proto-Costanoan/Proto-Miwok (i.e., Proto-Utian) comparison:

(1) 3663 (anchor date) × .29 = 1062.3
(2) 3663 + 1062.3 = 4725.3 (rounded to 4725)
(3) 3663 − 1062.3 = 2600.7 (rounded to 2601)
(4) 4725 to 2601 BP = range of dates in which the most likely breakup date for Proto-Utian is found (divergence date is not necessarily the anchor date)

The range can be interpreted with reference to the conclusions of Holman et al. (2011:853) from their fifty-two calibrations: "The mean absolute percent discrepancy is 29%; of the 52 ASJP dates, five are off by more than 50%, and one is off by more than 100%."

Table 9.1 supplies ASJP anchor dates and date ranges for twelve proto-languages. Anchor dates and ranges are not available for Proto-Aleut-Proto-Utian (aka Proto-Aleut-Utian) because the associated similarity indices are less than zero (negative). ASJP dates cannot be calculated from negative similarity indices. Nevertheless, from linguistic evidence other than specific ASJP dates (chapter 6) and archaeological and other evidence (9.4.1 and chapter 8), the most likely general time for the breakup of Proto-Aleut-Utian into Proto-Aleut and Proto-Utian was the Middle Holocene.

Table 9.1 ASJP Similarity Indices, Anchor Dates, and Date Ranges for Proto-Languages (of Pertinence to this Study) Ranked by ASJP Anchor Date from Largest to Smallest (Holman et al. 2011)

Language for Which ASJP Chronology is Estimated	*ASJP Similarity Index*	*ASJP Anchor Date*	*ASJP Date Range Based on 29 percent Discrepancy of Anchor Date*
Proto-Aleut-Proto-Utian	−0.53	NA (probably Middle Holocene)	NA
Proto-Salish	7.44	3827 BP	4937–2717 BP
Proto-Utian	8.29	3663 BP	4725–2601 BP
Proto-Wakashan	14.80	2781 BP	3588–1974 BP
Proto-Central Salish	18.29	2459 BP	3172–1746 BP
Proto-Miwok	22.54	2141BP	2762–1520 BP
Proto-Eskimo	27.43	1842 BP	2376–1308 BP
Proto-Numic	29.40	1737 BP	2241–1233 BP
Proto-Yupik	34.20	1506 BP	1943–1069 BP
Proto-Costanoan	39.23	1297 BP	1673–921 BP
Proto-South Wakashan	43.11	1154 BP	1489–819 BP
Proto-Inuit	60.40	640 BP	826–454 BP
Proto-North Wakashan	61.78	606 BP	782–430 BP

Source: Generated by the authors.

ASJP Chronology is one of several automated programs for dating the divergence of proto-languages. Among others, the most well-known is Bayesian dating (e.g., Gray and Atkinson 2003, Gray and Greenhill 2009). Citing several problems with the Bayesian approach, Rama and Wichmann (2020) develop a hybrid method, combining aspects of both ASJP and Bayesian, called Generalized Bayesian Dating (GBD). In the same paper, these researchers apply a method for evaluating different dating approaches for relative accuracy. Approaches evaluated include ASJP Chronology, Generalized Bayesian (first used in a publication by Barbieri et al. 2022), and a modified version of Swadesh's (1952, 1955a) Classic Glottochronology. Results are reported by Rama and Wichmann (2020:14) as follows: "Both in terms of mean absolute error with respect to 30 known dates, root mean squared error, and correlations between those dates and the inferred ones, the rank of performance is such that ASJP works better than GBD, which works better than glottochronology."

ASJP determination of proto-language divergence requires a classification of each language and dialect in its meta-data. Since its inception, ASJP Chronology has used the world-language classification provided by *Ethnologue,* a source updated over the years in twenty-five editions, the current edition being Eberhard, Simons, and Fennig (2022). As reported by Rama and Wichmann (2020), this classification is now being replaced in ASJP by the more conservative *Glottolog* (Hammarström et al. 2022). ASJP dates and similarity indices used in this study are based on *Glottolog*, unless indicated otherwise. We are grateful to Søren Wichmann for supplying needed materials to make the change from *Ethnologue* to *Glottolog* possible for this book.

9.3 CLASSIFICATION AND BORROWING

Figure 9.1 presents two family-tree classifications, one for Aleut-Utian languages and one for Eskimo languages, with ASJP date ranges for proto-language breakups. The trees are presented together to show graphically borrowing between languages of the two families and when this interaction occurred.

Proto-Aleut-Utian (PAU) is found at the highest branching of one of the trees. Before the branching, two different dialects pertained to PAU, one of which was to develop into Proto-Utian (PU) and the other of which became Early Aleut. This split of PAU was likely influenced by the migration of Utian-dialect speakers away from speakers of the Aleut dialect, thus isolating them from one another. Over time, this separation was instrumental in

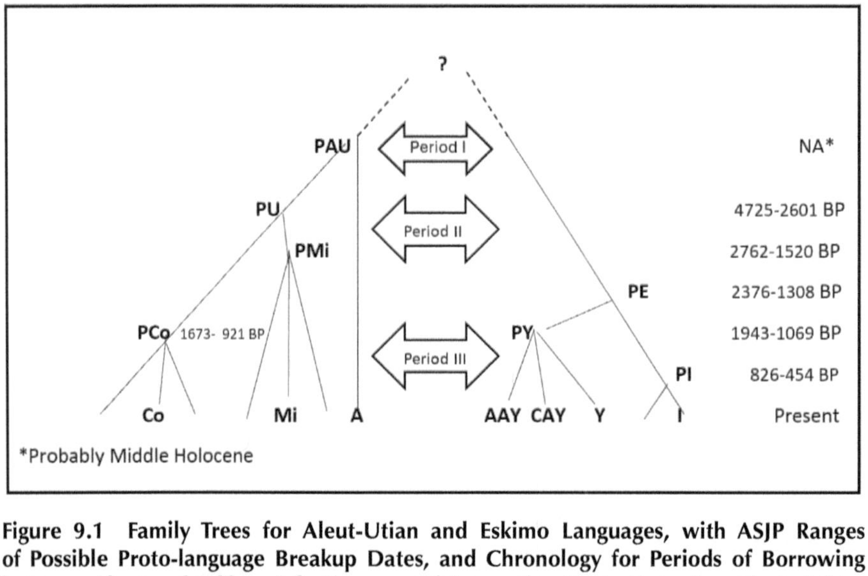

Figure 9.1 Family Trees for Aleut-Utian and Eskimo Languages, with ASJP Ranges of Possible Proto-language Breakup Dates, and Chronology for Periods of Borrowing between Aleut and Eskimo (cf., Berge, no date, section 3.1.). *Note*: A = Aleut; AAY = Alaska Alutiiq Yupik; CAY = Central Alaska Yupik; Co = Costanoan languages; I = Inuit languages; Mi = Miwok languages; PAU = Proto-Aleut-Utian; PCo = Proto-Costanoan; PE = Proto-Eskimo; PI = Proto-Inuit; PMi = Proto-Miwok; PU = Proto-Utian; PY = Proto-Yupik; Y = Other Yupik languages. Two-way arrows represent two-way lexical borrowing. Period I = borrowing between PAU and PE. Period II = borrowing between Aleut and PE. Period III = borrowing between Aleut and offspring of PE (i.e., PY, PI, Y, AAY, CAY, and I). *Source*: Generated by the authors.

the development of two distinct languages from these two dialects (9.5.1), that is, Proto-Utian and Early Aleut. While we do not know the duration of this process, we suspect the complete removal of these peoples from one another could have accelerated language separation substantially over the pace expected for the usual course of such divergence.

During or after the migration, some significant changes occurred in Proto-Utian phonology. Utian words were borrowed into some native languages of British Columbia and the Olympic Peninsula (Washington State) (chapter 6). Assuming they have been maintained mostly intact by the borrowing languages into modern times, these loans provide insight into the phonology of Proto-Utian at the time of borrowing. We dub Utian of this period as "Early Proto-Utian" in contrast to "Late Proto-Utian," the language spoken after its speakers settled in California, that is, the Proto-Utian reconstructed by Callaghan (2014) and used in this work to reconstruct Proto-Aleut-Utian (chapter 2, 2.4.6). The phonological changes from Early to Late Proto-Utian are as follows (see 6.7 for details):

Early PU **q* → Late PU **k*
Early PU glottalized consonants (e.g., **p'* and **k'*) → Late PU plain consonants (e.g., **p* and **k*)

The inclusion of Eskimo languages in figure 9.1 does not imply their genealogical relationship with any Aleut-Utian languages, which is why the two major left and right diagonal branches leaning toward one another do not fully connect at an apex. As discussed at length in 5.5, conventional thinking is that Aleut and Proto-Eskimo (PE) are sister branches of Proto-Eskimo-Aleut. These two languages show many grammatical and lexical similarities. The evidence we assemble in Chapter 5 indicates, however, that many, if not all, resemblances are due to borrowing rather than to language kinship. The question mark at the top of the tree (figure 9.1) suggests the inconclusiveness of the Aleut-Eskimo genealogical relationship.

The three double-headed arrows of figure 9.1 indicate periods when extensive borrowing between Eskimo languages and Aleut-Utian languages has taken place. Period I involves borrowing between PE and Proto-Aleut-Utian before PAU divided into Proto-Utian and Early Aleut; period II involves borrowing between PE and Aleut before PE split into Proto-Yupik (PY) and Proto-Inuit (PI) (2376–1308 BP); and period III involves borrowing between Aleut and offspring languages of PE (cf., Berge no date). The three periods taken together involve a chronological spread of several thousand years, during which Eskimo and PAU languages could have been in contact consistently off and on, if not continuously, and would have encompassed sufficient time for Aleut and Eskimo to develop substantial resemblance through borrowing.

When a lexical similarity is observed for Eskimo and PAU languages, determining in which of the three periods borrowing took place is relatively straightforward. If one of the two words is from PAU and the other from PE, borrowing occurred in period I; if one is from Aleut and the other from PE, borrowing is from period II; and if one is from Aleut and the other is from an offspring language of PE, borrowing is from period III. Examples (from Fortescue et al. 2010, except the PAU forms, which we reconstruct) are presented in table 9.2.

While it is relatively simple to identify the general time of borrowing, a definitive determination of what language contributed a specific loanword

Table. 9.2 Loanwords Exchanged Between PAU, Aleut, and Eskimo in Each of Three Different Time Periods

Period I Loan[1]	*Proto-Aleut-Utian Word*[2]	*Proto-Eskimo Word*
	8. **wanV-* SPIRIT	**anəʁnəʁ* "breath or spirit"
	10. **ṭayɪ-* MAN	**taʁu* "human being"
	13. **ṭot-* HEAR	**tučaʁ-* "hear or understand"
	24. **luṭɪ-* ONE	**ataʁuciʁ* "one"
	27. **kanə-* SNOT, PHLEGM	**kakkiɣ* "snot"
	29. **kVrɔ* FOOT	**kitmiɣ* "heel"
	43. **ʔekʊ* HAND, FINGER	**iqəlquʁ* "little finger"
	46. **ʔar-* BELOW, UNDERSIDE	**atliʁ* "lower thing," *at(ə)-* "down"
	47. **ʔay-* MORE	**ayu-* "go ahead or further"
	48. **ʔʊqu-* DOOR, GATE	**aqu* "stern"
	49. **line* AIR, SKY	**ini-* "hang out to dry"
	59. **imolV* HAIR OF HEAD	**məlquʁ* "body hair (or fur) or feather"
Period II Loan[3]	*Aleut Word*	*Proto-Eskimo Word*
	aɣlu-χ "killer whale"	**a(C)aałuɣ* "killer whale"
	akliɣa-χ "harpoon with bladder float"	**akləɣaʁ* "bird dart" (with bladder float)
	saami-χ "stone knife dagger"	**caviɣa* "knife"
	qyat- "tighten"	**əqət-* "squeeze"
	kayu-χ "strength, muscle"	**kayu* "be strong"
Period III Loan[3]	*Aleut Word*	*Proto-Eskimo Offspring Word*
	qakiiða-χ "silver salmon"	AAY *qakiiyaq* "silver salmon"
	qaðɣi- "penetrate"	PI **qažɣitaq* "pitfall trap"
	taaqa-χ "silverhake or walleye"	AAY *taaqaq* "tomcod"
	cuniɣ "backbone"	PI **tuniʁžuk* "breastbone"
	tumqi-χ "wooden pail"	PY **tuqmik* "container of some sort"

Source: Generated by the authors.
[1]This is a list of *all* period I borrowings we have been able to identify.
[2]The line number is the ID assigned to the PAU reconstruction in chapter 2, 2.4.6.
[3]Five loan pairs were selected from among many.

to another is difficult and often may not be possible. Establishing where languages were in contact when borrowing occurred is also problematic, but educated speculation based on the locations of proto-language homelands can help.

9.4 LANGUAGE HOMELANDS

In this section, we locate homelands for PAU, PE, EA, and PU. A "homeland" (or *Urheimat*) is the geographic region of a proto-language before the language split into offspring. Locating homelands is a basic step in reconstructing language history, and several different approaches for doing so are available (cf. Wichmann and Rama 2021, Wichmann et al. 2010). The strategy used here, correlating proto-language names for biological species with their geographic distributions, is among the most frequently employed. A problem with this approach, however, is that geographic ranges of modern biological species, most often the only distributional reports available, may have changed, sometimes significantly so, from those of proto-language times, depending on the chronological depth of the study.

9.4.1 Proto-Aleut-Utian Homeland

The vocabulary reconstructed for Proto-Aleut-Utian in chapter 2 (2.4.6) indicates a homeland for PAU speakers located in a coastal area. Of the sixty-six reconstructed words, 19, or 29 percent, designate maritime things. These nineteen words are given here with both their Aleut (A) and PU reflexes:

(Line numbers are those used in chapter 2 as IDs for each of the sixty-six comparative sets, 2.4.6.)
(*Note*: PU = Proto-Utian, PU[a] = authors' reconstruction of PU, → = "yields" or "develops into")

MARINE MAMMALS
38. PAU **qɔ-* SEA LION → A qawa-χ "sea lion," PU[a] **kota* "sea lion"
3. PAU **pa-* OTTER → A *halu[-]χ* "river otter," PU[a] **paw* "sea otter" (Aleut word is derived from Early Aleut **palu[-χ].*)
62. PAU **isowiq* HARBOR SEAL → A *isuχ* "harbor seal," PU[a] **so:-wik* "harbor seal"

MOLLUSK (SHELLFISH)
19. PAU **sɪw-* SHELLFISH SPECIES → A *sayɣi-χ* "mossy chiton," PU[a] **siwentak* "shellfish"

20. PAU *sar- MUSSEL → A *satmaayu-n* "big blue mussels," PU[a] **salo* "mussel"
35. PAU **quṭ-* RAZOR CLAM (*Siliqua patula*) → A *qatmi-χ* "razor clam," PU[a] **ku:ṭa* "Washington clam"

FISH
16. PAU **čawaC-* FISH TRAP → A *čaya-χ* "fish trap in river," PU[a] **čáwah* "fish net"
25. PAU **lʊk-* FISH LINE → A *akaɣðaχ* "fish line," PU[a] **ḷuk* "fish line"
31. PAU **qa-* FISH, COHO/SILVER SALMON (*Oncorhynchus kisutch*) → A *qa-χ* "fish," PU[a] **ka:ṣi* "salmon, coho"
57. PAU **hʊma* FISH EGGS (ROE) → A *ama-χ* "fish roe," PU[a] **hu:ma* "salmon eggs, fish eggs"

OTHER MARINE CREATURE
51a. PAU **ʔorʌ* SEA ANEMONE (order Actiniaria) → A *utu-χ* "sea anemones," PU[a] **ʔo:la* "sea anemone"

MARINE PLANTS
18. PAU **čɔ(qa)l-* SEAWEED → A *čaqalka-n* "kind of seaweed," PU[a] **čol:* "kelp, seaweed"
32. PAU **qa-* KELP, SEAWEED → A *qaŋu-χ* "seaweed, kelp," PU **ka-* "type of seaweed"

MARINE ENVIRONMENT
22. PAU **ṣolo-* ROAR OF OCEAN → A *sulux-six* "to roar (of ocean)," PU **ṣolo* ~ **čolo* "to roar (river)"
26. PAU **ka(m)læ-* WAVE(S) → A *kamla-l* "to form swells or large waves," PU[a] **kalen* "wave"
45. PAU **ʔalə-* WAVE(S) → A *aalu-χ* "swell, big wave," PU[a] **ʔa:len* "wave(s)"
53. PAU **ʔɔlɔʁ-* OCEAN, SEA → A *alaʁu-χ* "the sea, the ocean," PU[a] **ʔo:lok* "ocean"

WATERCRAFT
21b. PAU **ṣɨka-* BOAT → A *suka-χ* "baidarka skirt," PU[a] **ṣak(:)as* "boat"
50a. PAU **ʔolaʁ* HATCH OF BOAT, OAR, ROW → A *uluʁu-χ* "hatch of boat," PU[a] **ʔo:lak* "oar, to row"

While referents of these PAU words unambiguously indicate a coastal homeland for the proto-language, none directly suggest where that coast might have been. Locations of PAU daughter languages indicate two likely general prospects: coastal California (Proto-Utian) and coastal Alaska (Aleut). Since the known history of the peopling of the Americas involves a general north-to-south movement, ancient migration of groups from California to Alaska seems implausible. Consequently, we assume PAU's coastal homeland was in Alaska.

Three major bodies of water define Alaska's north, west, and south borders, respectively, the Arctic Ocean, the Bering Sea, and the Gulf of Alaska (figure 2.1). All three marine mammals designated in the above list of PAU terms (sea lion, sea otter, and harbor seal) are found in the Gulf of Alaska. None of the three are found in the Arctic Sea, and two of the three, harbor seal and sea otter, in addition to the Gulf of Alaska, inhabit waters of the Aleutian Islands. The sea lion is found throughout the Bering Sea as well as in the Gulf of Alaska and the Aleutians. The razor clam *Siliqua patula* (PAU **quṭ-*) has a similar distribution to that of the harbor seal and the sea otter (found in the Gulf of Alaska and Aleutian waters, but not in the Bering Sea or the Arctic Ocean) (https://www.adfg.alaska.gov).

Other evidence for this homeland is the distribution of several common marine mammals of Alaska not found in the Gulf of Alaska or the Aleutians that, as might be expected, do not have reconstructable PAU names. These mammals are bearded seal, ribbon seal, ringed seal, spotted seal, bowhead whale, and walrus, words for all six of which reconstruct for Proto-Eskimo; PE words do not reconstruct for sea lion, sea otter, harbor seal, and razor clam (Fortescue et al. 2010), indicating a PE homeland somewhere other than along the shores of the Gulf of Alaska (9.4.2).

Given these observations, the PAU homeland could have been situated anywhere along the Gulf of Alaska coast, from the Aleutians in the west, moving eastward around the coast of the Alaska Peninsula, into the greater Kodiak Archipelago, around the southern reaches of the Kenai Peninsula, and ultimately to the eastern shores of south Alaska. Of these possible locations, a PAU homeland in the Kodiak area seems a reasonable conjecture (figure 10.1).

Kodiak Island was uninhabited until the Middle Holocene (Fitzhugh 2004:18). Its first colonists were dependent on marine resources, including large mammals. Their complex maritime tradition, called Ocean Bay (8.4.1), is documented archaeologically as enduring from 7500 to 3500 BP (Fitzhugh 2003a). If PAU speakers were the first group to colonize Kodiak Island and abutting areas, this would have taken place around the beginning of the Middle Holocene (8200 to 4200 BP), and thus the migration of our ancient mariners to San Francisco Bay would have been a Middle Holocene event.

The greater Kodiak region is geopositioned for taking advantage of coast-hugging routes to destinations to the east and south, such as coastal British Columbia, the Olympic Peninsula, and the San Francisco Bay Area (chapter 6). Launching a strictly coastal voyage from other locales in southwest Alaska, especially from a more westerly location such as Aleutian waters, would not be so convenient, adding hundreds of kilometers (depending on launch location) to the trip. On the other hand, taking a shorter direct route would require navigation across open sea, out of sight of land for extended periods of time, during which our ancient mariners would have had to deal with increased threat situations such as violent storms with no option of quickly accessed coastal refuge. In brief, if our ancient mariners began their journey somewhere in southwest Alaska, it seems a more practical place for doing so would have been the greater Kodiak region rather than the Aleutians. (This paragraph assumes our ancient mariners knew where they were going. This possibility is discussed in 9.5.6 and 9.6.1.2.)

9.4.2 Proto-Eskimo Homeland (in Alaska)

Identifying the PAU homeland helps to nail down where Proto-Eskimo might have been spoken when its speakers first settled in Alaska. PAU and, later, its daughter language Early Aleut must have been in close geographic contact with PE, given the considerable size of the borrowed vocabulary PE shares with these two languages (5.5 and 9.3).

Scholars have diverse opinions on where Proto-Eskimo was spoken (cf., Dumond 1984:78–79), some suggesting locations in Alaska, others speculating on earlier locations in far Eastern Asia across the Bering Sea (Anna Berge, pers. com. 2021). Berge conjectures that the PE homeland "is either in the area around St. Lawrence Island and the Chukchi Peninsula, or the area around the Alaska Peninsula; it depends on how far back we go and which pieces of evidence we use. Whaling appears first in Southwest Alaska, but the distinctive Eskimo culture appears to emerge first in far Eastern Asia, quite late." Despite these differing opinions, all seem to agree on the presence of PE in ancient Alaska at one time or another, and this understanding informs our reconstruction.

Table 9.3 presents words from Proto-Eskimo for eighteen zoological species. Geographic ranges in Alaska for these creatures are given on the website of the Alaska Department of Fish and Game. Distribution maps or discussions of distributions are presented on the website for each of 156 creatures. Among the 156, only the eighteen species listed in table 9.3 have Proto-Eskimo names reconstructed by Fortescue et al. (2010) and published in their massive *Comparative Eskimo Dictionary*. Also presented in table 9.3 are Aleut words for eleven of the eighteen species (these are from Bergsland's similarly massive *Aleut Dictionary*).

The range across Alaska is indicated for each of the species listed in table 9.3. Each occurs (+) or does not occur (−) in each of ten areas (Kodiak Island, Aleutian Archipelago, Seward Peninsula, etc.). Figure 9.2 presents a map of Alaska with the ten regions roughly identified. The number of the eighteen

Table 9.3 Distribution across ten areas of Alaska of 18 zoological species for each of which a Proto-Eskimo name reconstructs (Fortescue et al. 2010), and an Aleut name if found in Bergsland's (1994) dictionary

Zoological Species: Proto-Eskimo (PE) label / Aleut (A) label	*Ten Regions of Alaska (See Names Matching Numbers Below)*									
	1	*2*	*3*	*4*	*5*	*6*	*7*	*8*	*9*	*10*
River Otter: PE: *piʁtuʁčiʁaʁ / A: halu[-]χ	+	+	+	+	+	+	+	+	+	+
Common Raven: PE: *tulukaʁ / A: qalŋaaχ	+	+	+	+	+	+	+	+	+	+
Ringed Seal: PE: *nayyiʁ / A: puchaaskigî-lix	+	+	+	+	+	+	+	+	–	–
Caribou: PE : *tuntu / A : itχayax	+	+	+	+	+	+	+	+	–	+
Wolverine: PE: *qatviɣ / A: qačimayugna-χ	+	+	+	+	+	+	+	+	–	+
Muskrat: PE: *kəvɣaluɣ / A: no word	+	+	+	+	+	+	+	+	–	+
Moose: PE: *tuntuvaɣ / A: no word	+	+	+	+	+	+	+	+	–	+
Lemming: PE: *avəlŋaʁ / A: anamču-χ, asχukiδ-χ	+	+	+	+	+	+	+	+	–	–
Walrus: PE: *ayvəʁ / A: amɣaaδa-χ	+	+	+	+	+	+	+	+	–	–
Killer Whale: PE: *č(C)aʁluy / A: aɣlu-χ, aʁaʁiya-χ	+	+	+	+	+	+	+	–	+	+
Bearded Seal: PE: *maklaɣ / A: no word	+	+	+	+	+	+	+	–	–	–
Beluga Whale: PE: *čitu(C)aʁ / A: xaδa[-]χ, miluuɣa-χ	+	+	+	+	+	+	–	–	–	+
Arctic Ground Squirrel: PE: *čikδiɣ / A: no word	+	+	+	+	+	–	+	+	+	–
Arctic Fox: PE: *təʁikanni(C)aʁ / A: no word	+	+	+	–	+	+	–	–	–	–
Humpback Whale: PE: *qipuqqaʁ /A: alamax	–	+	–	+	–	+	+	+	+	+
Arctic Char: PE: *iqaluɣpiɣ / A: no word	+	+	–	+	–	–	+	–	+	+
Beaver: PE: kəɣi(C)aʁ / A: aŋtux, aaquya-χ	+	–	+	+	–	–	–	–	+	+
Polar Bear: PE: *nanuʁ / A: no word	+	+	+	–	+	–	–	–	–	–
TOTALS FOR + (PLUS SIGN)	17	17	16	16	15	14	14	11	7	11

Source: Generated by the authors.

Notes: Names of numbered regions: 1. North Slope, 2. Seward Peninsula, 3. Selawik Lake Region, 4. Bristol Bay Area, 5. Point Hope Area, 6. Hooper Bay Area, 7. Alaska Peninsula, 8. Aleutian Archipelago, 9. Kodiak Island, 10. Kenai Peninsula.

species found in each area is given (information extracted from table 9.3). For example, seventeen of the eighteen occur on both the North Slope and the Seward Peninsula, and only seven on Kodiak Island. No single region shows PE words for all eighteen species.

The map in figure 9.2 shows a continuous distribution of areas with from fourteen to seventeen of the eighteen creatures. These areas include, from south to north, the Alaska Peninsula (14), the Bristol Bay Area (16), the Hooper Bay Area (14), the Seward Peninsula (17), the Selawik Lake region (16), the Point Hope area (15), and the North Slope (17). Any area within this high-number zone plausibly could have been the PE homeland.

The map's remaining areas show substantially lower numbers of the eighteen species, that is, from seven to eleven. These are areas located in coastal southwest Alaska: Aleutian Archipelago (11), Kodiak Island (7), and Kenai Peninsula (11). If the extreme southwestern end of the Alaska Peninsula is considered separately from the rest of the peninsula, we can include it with the former three low-number areas since it shows only seven of the eighteen creatures. By adding this area, the low-number region shows a more or less continuous distribution across coastal south Alaska, from the Aleutians to the Kenai Peninsula. The location of the Proto-Eskimo homeland in this region is unlikely compared to the more northern regions with higher numbers of PE-named species.

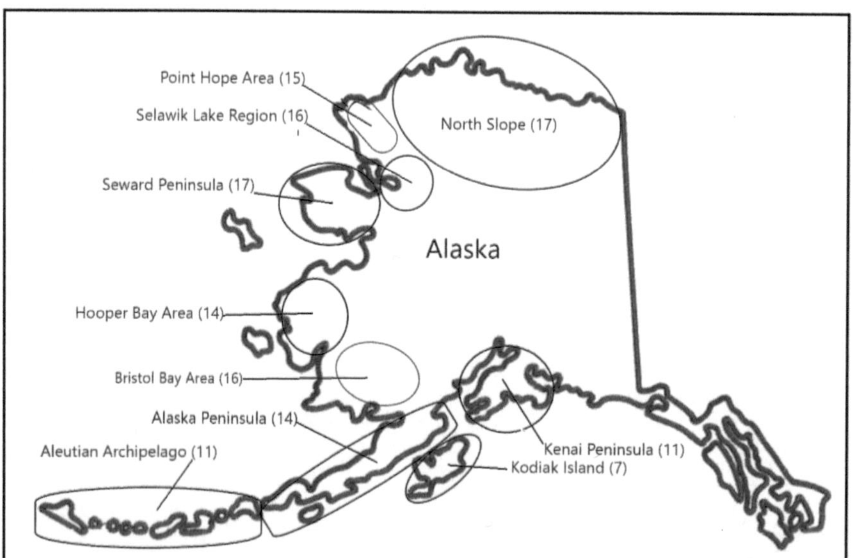

Figure 9.2 Map Locating the Ten Areas of Alaska Cited in Table 9.3 Showing the Number Found in Each Area of the Eighteen Zoological Species with a Name Reconstructing for Proto-Eskimo. *Source*: Generated by the authors.

The higher-number species area of Alaska (14–17) mostly corresponds with those regions in which modern-day Eskimo languages (Yupik and Inuit) are spoken (except for the extension of Pacific Yupik Eskimo into the Kodiak-Kenai area). The lower-number area of Alaska (7–11) corresponds both to where present-day Aleut is spoken (Aleutian Archipelago) and to where we propose its ancient ancestor (Proto-Aleut-Utian) was spoken (see figure 10.1 and section 9.4.1).

The high-number area comprises those parts of Alaska where the Proto-Eskimo homeland could have been located, including the Alaska Peninsula (14 species), which is geographically adjacent to the greater Kodiak region, the low-number area we propose as the PAU homeland (figure 9.2). The conjoined Alaska Peninsula and Kodiak areas, then, constitute the geographic zone of Alaska in which speakers of PE and of PAU were in closest proximity and is the likely area in which these languages exchanged loanwords (during period I of borrowing; see figure 9.1).

9.4.3 Early Aleut Homeland (in Alaska)

The breakup of PAU produced two daughter languages, Proto-Utian and Early Aleut. Utian migration away from the greater Kodiak area could have resulted in Early Aleuts as the main proprietors of what was once the PAU homeland. (Both Berge [2018, 2024], a linguist, and Maschner [2016], an archaeologist, suggest Kodiak culture [Ocean Bay Tradition] was originally Aleut.) If borrowing between Aleut and Proto-Eskimo is any indication (during period II, see figure 9.1), Early Aleuts occupied the old PAU homeland for a considerable period of time, and their interaction with PE speakers could have continued the loanword exchange begun by speakers of PE and PAU (in period I).

Prior to the Proto-Eskimo breakup, speakers of the language began to move northward from the Alaska Peninsula along the west coast of Alaska, ultimately to the North Slope and beyond to the east. This could have diminished or even emptied the Alaska Peninsula of most Eskimo speakers, allowing Aleuts to move relatively unimpeded from their homeland region into the area, from which they subsequently migrated down the peninsula into the Aleutian Archipelago.

9.4.4 Proto-Utian Homeland (in California)

Our reconstruction of ancient events proposes Utian colonists came to central California via a water route, contradicting the view of Utian migration from some interior area outside of California, notably the Great Basin (Moratto 1984:552ff) but also the Columbia Plateau. Moratto's story has Utians settling at the Windmiller area near modern Stockton, California. This is about

130 km due east of the San Francisco Bay Area, SFBA being our proposed location for the Proto-Utian homeland in California. The ASJP date range for Proto-Utian is 4725–2601 BP, with an anchor date of 3663 BP. (Mitochondrial DNA evidence assembled in chapter 4 [4.4] accords with our location of the Proto-Utian homeland in the San Francisco Bay Area.)

Callaghan (2014:24) cites James A. Bennyhoff and David Frederickson's personal communication that the "Lower Berkeley Pattern [archaeological tradition], located inland from the Bay Area and dated around 4000–4500 years ago, represented Proto-Utian speakers." For this proposal, she (2014:24) supplies some weak linguistic evidence, writing that this location is consistent with Proto-Utian terms for three plants: poison oak, gray pine, and valley oak. She adds, however, "Of course, these plants are found in many other parts of California as well."

Callaghan cites another personal communication from Frederickson proposing the North Bay as the homeland of Proto-Miwok (one of the two daughter languages of Proto-Utian) with Miwok "subsequently spread west to the Coast." According to Callaghan (p. 24), speakers of Proto-Costanoan (the other Proto-Utian daughter language) settled in a homeland in the East Bay Area. She also mentions Levy's (1978:2–3) observation "that many Coast Miwok and Costanoan terms for plants and animals peculiar to the coastal ecosystem are borrowings [rather than inherited words], which supplies supporting evidence against a coastal homeland for the Proto Utians." Unfortunately, we were unable to find copies of Levy's paper (or papers) despite extensive efforts to obtain them. We have also been unable to identify the linguistic borrowings to which he has apparently referred.

Our narrative is very different. In the Middle Holocene, Utian migrants navigated to and made landfall in the San Francisco Bay Area, bringing with them their maritime way of life. There they exploited a wide range of marine resources. Sometime within the ASJP date range of 4725–2601 BP, Proto-Utian divided into the two daughter languages, Proto-Miwok and Proto-Costanoan (figure 9.1). This division was likely a result of the spread of Utians from their original SFBA homeland into northern (Proto-Miwok) and southern (Proto-Costanoan) Bay areas.

Subsequently, both Proto-Miwok and Proto-Costanoan began to diverge into their respective daughter languages (figure 9.1), when some speakers of Miwok left the Bay area, moving into the interior, and when Costanoan speakers became distributed along the sea coast south of the Bay. For Miwoks, this migration involved following major rivers and their tributaries upstream, with some groups reaching as far east as the Sierra Nevada. These people abandoned their maritime way of life, adopting the subsistence strategy of California Natives focused on acorns (7.8). Other groups of Miwoks remained behind in the west, near to or in the Bay area, continuing to make a living by harvesting food from the bays and sea. (See figure 2.2 for a geographic distribution of Utian languages in the historic era.)

If Proto-Utians in California had a maritime lifestyle in a coastal homeland, reconstructions for the language should include words for marine plants and animals. In her Proto-Utian dictionary, however, Callaghan (2014) only reconstructs one PU word for a marine species. This is **has:ɨn* "abalone, abalone shell," with reflexes in Costanoan (Csjb *haš:an* "abalone, abalone shell") and Miwok (PMis **has:ɨn* "abalone") languages (see sections 2.2.2, 2.4.1, 2.3, or figure 9.1 for Utian-language abbreviations). Although some might argue that reconstruction of just one PU word for a marine plant or animal is sufficient evidence for positing a PU coastal homeland, Callaghan does not make such an argument, and neither do we.

Callaghan reconstructs a PU word only if one reflex occurs in a Costanoan language and another one in a Miwok language (2.1). We use a different, somewhat less conservative approach for our PU reconstructions (identified by PU[a]). If a Miwok or Costanoan word (but not both) and an Aleut (A) word appear to be cognate, we regard this as evidence for the reconstruction of both a PAU word and a Proto-Utian word (if sound correspondences are observed). For example, the following is a reconstruction for PAU with only a Costanoan term and an Aleut word as supporting evidence:

Cscr [*saro*] "mussel"
A *satmaayu-n* "big blue mussels"
PAU **sat-* MUSSEL → PU[a] **salo* "mussel"

And another example with only a Miwok term and an Aleut word as supporting evidence:

Mib *ʔóola* "sea anemone"
A *utu-χ* "sea anemones, Actinaria"
PAU **ʔota* SEA ANEMONE → PU[a] **ʔo:la* "sea anemone"

Using this approach combined with that of Callaghan, we find twelve words for marine species reconstructable to Proto-Utian (ID numbers are consistent with those used for sets in chapter 2, 2.4.6):

PU **has:ɨn* "abalone" (Csjb and PMis) (← PAU not observed)
3. PU[a] **paw* "sea otter" (Mim and A) (← PAU **pa-*)
18. PU[a] **čol:* "kelp, seaweed" (Mib/Mim and A) (← PAU **čɔ(qa)l-*)
19. PU[a] **siwentak* "shellfish" (Mim and A) (← PAU **sɪw-*)
20. PU[a] **salo* "mussel" (Cscr and A) (← PAU **sat-*)
31. PU[a] **ka:si* "coho, silver salmon" (Mim and A) (← PAU **qa-*)

32. PU[a] *ka-* "type of seaweed" (Mib and A) (← PAU *qa-)
35. PU[a] *ku:ta* "Washington clam" (← Early PU[a] *qu:ta* "Washington clam") (Mib/Mim and A) (← PAU *quṭ-)
38. PU[a] *kota* "California sea lion" (Mim and A) (← PAU *qɔCa-)
51a. PU[a] *ʔo:la* "sea anemone" (Mib and A) (← PAU *ʔota)
57. PU[a] *hu:ma* "salmon eggs, fish eggs" (Mib and A) (← PAU *hʊma)
62. PU[a] *so:-wik* "harbor seal" (Cru and A) (← PAU *isowiq)

We submit these lexical reconstructions as confirmation of a California homeland (in SFBA) for speakers of Proto-Utian.

With the exception of PMis (Proto-Sierra Miwok), all of the Miwok (Mib and Mim) and Costanoan (Csjb, Cscr, and Cru) languages supporting the twelve above-listed reconstructions are coastal languages. All other Miwok languages (Miss, Mics, Mins, Mip, and Mil) show very few or no words for marine plants and animals since speakers of Miwok dialects ancestral to these languages moved away from the coast to the central California interior before Proto-Miwok broke up into daughter languages (ASJP 2783–1531 BP). These interior Miwok languages lost words for marine plants and animals simply because they did not find these flora and fauna in their adopted interior habitats (cf., Brown 1981, 1984:46–47).

Callaghan's reconstruction of PU *has:ɨn* "abalone, abalone shell" does not fit this pattern. The supporting Miwok word for her reconstruction is PMis *has:ɨn* "abalone, abalone shell" based on Miss *has:ɨn* "abalone shell' and Mics *has:ɨn* "abalone." While Miss and Mics are interior languages and abalone are marine creatures that do not naturally occur in interior habitats, abalone shells may have been common trade items between coastal and interior Miwok people. The inner layers of the abalone shell are mother-of-pearl, which is widely used in jewelry and in other decorative ways.

Speakers of Miwok languages migrated into the interior of central California, in some instances as far east as the Sierra Nevada eastern slopes. Such migration would put some Miwok groups in relative proximity to Numic (Uto-Aztecan) groups in the Great Basin area of Nevada State and abutting areas, such as speakers of Northern Paiute. Numic people expanded into the Great Basin relatively recently, as indicated by the ASJP date-range for Proto-Numic (2241–1233 BP, table 9.1) (cf., Lamb 1958:98–99).

Mitochondrial DNA evidence (chapter 4) indicates some Utian and Numic groups had more than just a casual acquaintance. Northern Paiute speakers, for example, share with Utian speakers high-frequency *D* haplogroups (most-frequent *D*), a relatively rare feature among Native American populations (table 4.1, figure 4). In addition, Northern Paiute and Yukisma SCL-38 2 (an ancient site of the San Francisco Bay Area) have identical relative-frequency profiles (comparative score 4). This close similarity among Great Basin populations could be due to admixture (5.5), with Numic-language speakers marrying Utian speakers. In addition, language shift is another possible convergence factor. Morgan (2010:164, 168) discusses evidence for frequent intermarriage between Mono (Numic) and Miwok speakers on the Southern Sierra Nevada eastern slopes.

9.5 THE UTIAN MIGRATION

The following are questions about the Utian migration and our current-best, although sometimes only tentative, answers (reconstructions):

9.5.1 Who Were the Ancient Mariners?

Ancient mariners who undertook their voyage (or voyages) from southwest Alaska to central California were marine food harvesters of the Ocean Bay tradition (8.4.1). Maritime culture, such as that of our migrants, is described by Workman and McCartney (1998:361–362) as entailing "human groups which live in proximity to, and obtain a significant portion of their subsistence from, the sea . . . maritime cultures possess(ed) seaworthy watercraft and obtained at least some of their caloric income from resources procured beyond the littoral zone."

In preceding chapters, evidence is discussed for identifying our ancient mariners as speakers of Proto-Aleut-Utian. This includes reconstructed PAU vocabulary for words naming maritime artifacts such as boats, boat parts, and organisms such as marine mammals. In chapter 2, we reconstruct sixty-six words for PAU (2.4.6), nineteen of which designate items relating to a marine environment. These nineteen words are grouped together and listed in 9.4.1.

Speakers of Proto-Utian or of a dialect of PAU developing into Proto-Utian took the maritime way of life to central California from southwest Alaska. Proto-Utian also was clearly a maritime language since it retained all the maritime words of its parent language.

9.5.2 When Did the Migration Take Place?

Utian migration to California probably began sometime in the Middle Holocene (8200–4200 BP).

9.5.3 Why Did They Migrate?

Anthony (1990) describes this question as one of the most difficult in archaeological migration studies.

A possible answer is discussed in chapter 7, 7.3.1: Cooler weather associated with the Neoglacial of the Middle Holocene may have motivated population movements and could have been the impetus for the Utian southward migration.

9.5.4 How Did They Travel?

Reconstructed Proto-Utian words for nineteen marine referents in 9.4.1 indicate retention of a strong maritime focus by our migrants as they journeyed by watercraft from Alaska to California. This emphasis might not have been maintained if they had instead chosen an interior route. But why people equipped with the technology and skills of seafaring might prefer an inland route over an obvious marine route would be difficult to explain.

Two of the nineteen PAU words are especially indicative of seafaring knowledge and maritime culture: 21b. **ṣɨka-* BOAT, BAIDARKA SKIRT, a word retained in Late PU as **ṣak(:)as* "boat" and 50a. **ʔolaʁ*—HATCH OF BOAT, OAR, ROW, retained in Late PU as **ʔo:lak* "oar, to row." While we do not know for certain what types of boats were built, this lexical reconstruction is a near "smoking gun" attestation of PAU and Proto-Utian use of watercraft. Thousands of years of decay of wood, bone, skins, and other organic materials from which boats probably were constructed would preclude any possibility of finding intact or partially preserved Utian boats at Middle Holocene archaeological sites, but we are still hopeful some will be located. Our current best chance of learning what these boats were like is from the ethnohistorical record regarding watercraft of Indigenous Peoples of the North Pacific observed when Europeans first arrived.

An excellent review of historical Indigenous boats is Anichtchenko (2012). Two kinds of boats are discussed, the kayak and open-skin boats (umiak). Kayaks were traditionally constructed to accommodate no more than one or two people (cf. Wilson 2015:15). In contrast, open-skin boats (which are no longer in use) could be built to accommodate as many as seventy passengers and crew, and there are reports of Central Aleutian boats 10 to 12 meters long. Typically, large open-skin boats would have accommodated around twenty people. Fleets of open-skin boats are also reported, for example, a Koryak (Kamchatka) group of eighty boats navigated to the New Islands of the Aleutians. Because of their small size, kayaks were an unlikely watercraft to move large groups of people over thousands of kilometers, open-skin boats were more likely used by our mariners on their journey. (Open-skin boats are not observed for ethnohistorical groups of California, suggesting these were replaced in Utian coastal areas by traditional native California boats, the tule craft, and hardwood canoes.)

Historically, open-skin boats were propelled by paddlers, although these were replaced by rowers with oars and oarlocks under Russian influence (Anichtchenko 2012). In ancient times, some boats were possibly equipped with sails, although the evidence for this is equivocal (Anichtchenko 2013). Nevertheless, Fortescue et al. (2010:372) are able to reconstruct a Proto-Eskimo word for "sail," **təŋəlʁaʁ*. (The ASJP date range for PE is 2494–1372 BP.)

(An anonymous reviewer suggests other plausible interpretations of some aspects of Utian migration. According to the reviewer, Native American migrations [terrestrial] typically involve small groups of people traveling fairly rapidly in stages but not all at once. The reviewer sees Utian loanwords into languages encountered along the way as suggestive of a similar process: ancient mariners ever moving along the coast but stopping for one to several generations before continuing their journey. Utian people left behind after prolonged stopovers would have become assimilated to the lifestyles and languages of local groups.)

9.5.5 How Many People Migrated?

To estimate how large our migrant population must have been to survive once it had arrived and settled in central California, a key metric in conservation biology is used: the minimum viable population (MVP) size. In brief, MVP is the smallest possible size required for a biological population to resist extinction from various factors such as natural disasters and inbreeding. Browsing the literature yields a wide range of different MVP values for human populations, ranging from as small as 12 to 40,000 or even greater. The most compelling result does not consider human populations but rather is a compilation and statistical analysis of MVP findings for 212 species in the categories of birds, fish, mammals, herptiles, plants, insects, and marine invertebrates, based on studies published since the early 1970s (Traill et al. 2007). The average MVP for all 212 species after using a correction factor for standardizing differing approaches for different species is 4,169 (with a subaverage of 3,876 for mammals). The study's authors write of this result in their abstract: "our findings are consistent with biological theory and MVPs derived from abundance time series in that the MVP for most species will exceed a few thousand individuals."

Whether an MVP for human populations would "exceed a few thousand individuals" is unknown, but we considered this possibility in estimating how many Utian peoples moved from southwest Alaska to central California to produce a viable population. We then use this result to estimate how many open-skin boats, if more than one, were required for the migration endeavor. Our conservative guess is an MVP for humans of 500 individuals, or slightly more than 10 percent of our estimated size of the PAU population at the time of the migration, which is 4500 people. This estimate is based on a determination by Townsend (1981:637) of 4000 to 5000 Dena'ina (Athapaskan) at the time of European contact. The Dena'ina then and now inhabit an area (around Cook Inlet) close to the PAU homeland (figure 10.1).

The largest historically documented open-skin boat noted in the Anichtchenko review (2012) could accommodate seventy individuals. Assuming our mariners would have used the largest boats possible, at least seven watercraft would be needed to take 500 people to California. All seven boats departing Alaska's shores at approximately the same time would constitute a "fleet," in accord with historically observed fleets, one of which is cited in Anichtchenko's review (2012) involving a Koryak fleet of eighty open-skin boats journeying to the New Islands. Plausibly, instead of one fleet of seven boats, two fleets of four boats each could have achieved the goal of transporting at least 500 people to their destination. These two fleets could have been launched at different times, perhaps even a year or more apart, which would have reduced the pressures of launching and managing seven large boats and 500 people simultaneously. Some linguistic evidence for multiple fleet use for the trip to California is discussed presently (9.5.6).

The size of a viable Utian population that traveled to California may also have been influenced by voyagers' interactions with the resident population of the greater San Francisco Bay Area. A major observation of the archaeological analysis in chapter 8 is that Utians forged close relationships with resident communities that led to the merging of extant Native California people and Utian mariners. Smaller numbers of Utians may have trekked southward to California and produced a viable population by integrating with people who had inhabited the area for thousands of years.

9.5.6 Which Way to Go?

How did our ancient mariners know which way to go when their migration was launched? Plausibly, they had a plan based on information gathered by PAU mariners or other navigators, scouts (9.6.1.2), who had already traveled to their possible destination, since people of the circum-polar area may have regularly plied the great northern oceans for reconnoitering purposes in the Middle Holocene or earlier, typically leaving no or very little evidence that they had ever done so.

PAU speakers interested in migration to and colonization of central California may have invested considerable time and resources preparing for the migration, including the construction of large open-skin vessels (9.5.4) and stocking them with provisions, spare parts, hunting and fishing equipment, and trade items, in preparation for their journey. It seems unlikely such extensive efforts would have been undertaken based solely on a hunch about where they were headed.

When departing Alaska, migrating mariners followed the coast closely, keeping an eye dedicated to the "kelp highway," just as ancient mariners of the North Pacific might have done for thousands of years before them (Erlandson 2001). This route is suggested by PAU and Proto-Utian words reconstructed for "kelp" and "seaweed" (9.4.1). Hugging the coast would have allowed for easy, brief stops when provisions such as fresh water needed replenishing or watercraft required small repairs.

One indication of a coastal route for our mariners are loanwords from Utian borrowed into four languages of two different areas of the coastal Pacific Northwest (6.5). These are Haisla and Heiltsuk (two closely related North Wakashan languages) of coastal British Columbia, and Upper Chehalis (Salish) and Quileute (Chimakuan) of the northwest of Washington State. These two areas and associated languages are identified by two circles on the map of figure 6.1 (chapter 6). The areas are roughly 850 km apart as the crow flies.

Words borrowed from Utian by these languages indicate Utian speakers were in contact with coastal peoples for extended periods of time, probably due to prolonged layovers for wintering, boat maintenance, reprovisioning, trading, advanced scouting operations (9.6.1.2), and other reasons. While a single fleet of vessels could have had long visits at each of the two venues, a competing, simpler explanation is that two or more different fleets of migrants from southwest Alaska each had separate prolonged stayovers at different times, in different coastal locales.

In addition to loanwords, there is genetic evidence for prolonged interaction with a Native People of Middle Holocene British Columbia. The mtDNA haplogroup profile of speakers of Splatsin, a modern Salish language of southern British Columbia, shows a feature indicative of genetic admixture with Utian. This is the high-frequency *D* haplogroup in their genetic profile (chapter 4, 4.4). Although Splatsin territory is not exactly coastal, since it is located about 375 km from coastal waters, it is close to small rivers draining into the Pacific Ocean from the coastal mountains in southern British Columbia. Plausibly, Salish ancestors of Splatsin people may have encountered Utian mariners somewhere in a coastal area before moving to the interior, their current location.

9.6 MIGRATION AS PROCESS

In concluding this chapter, we consider how what we know about the Alaska-to-California journey fits the general structure of migration as discussed by the anthropologist, David W. Anthony (1990). Anthony's comments are chiefly directed toward archaeologists whose research in some manner involves migration, advising them to approach the topic bearing in mind that migration is not merely an event to be measured and described but rather a process of patterned human behavior. This may be a valuable approach, not only for the study of migration by archaeologists but also by historical linguists. Here we briefly examine how migration as a process, as discussed by Anthony (1990:899–905), can contribute to our investigation of the Utian migration from southwest Alaska to central California.

9.6.1 The Structure of Migration

9.6.1.1 Conditions Favoring Migration

Specific motivations for migrations are often difficult to nail down, which is certainly the case for the Utian migration (Anthony 1990:199; only page-number citations are given for this source hereafter). However, all migrations involve the basic formulation of the "push-pull model." In this model, negative stresses (push away from homeland) and positive attractions (pull toward the destination) are evaluated, looking at which is more acceptable, the cost of moving as opposed to just staying at home. For Utian migration, an important push factor might have been the climate, since Alaskan winters are such that marine food procurement is all but shut down due to frozen bays and inlets, and an obvious pull factor is the far warmer climate of central California. Another pull factor might be advances in boat-building technology supporting extensive hunting and fishing on the high sea, which could be the original rationale for building seacraft capable of travel over very long distances. Moreover, California has long held an attraction for many different groups of people of different origins and eras (ancient, historic, and modern).

9.6.1.2 Scouts

In long-distance migration, the activity of scouts is crucial since they survey areas possibly suitable as destinations and report this information back to potential migrants (pp. 902–903). One reason migrants envisioned coastal California as a desirable destination could have been favorable reports of returning exploratory mariners (scouts) regarding conditions in that region. From loanword evidence (6.6), however, we know of prolonged visits of Utians with Native Peoples of British Columbia (Wakashan) and northwest Washington State (Salish and Chimakuan). These two locations may have been planned stops on the way to California based on information from scouting mariners, or one or both locations could have been planned final destinations later abandoned by migrants after receiving information provided by scouts on the exceptional suitability of coastal central California.

9.6.1.3 Migration Streams

Citing Lee (1966:54), Anthony (p. 903) writes, "Migrants tend to proceed along well-defined routes toward specific destinations." These routes have often been recognized (if not sometimes blazed) by earlier migrants who have passed on routing information to later migrants. Despite the roughly 5000 km between coastal southwest Alaska and the San Francisco Bay area, a simple message regarding their migration to California might have been conveyed to our potential Utian colonists: "Just follow the kelp highway."

Citing Lefferts (1977:39), Anthony (p. 903) writes,

> In addition to following a well-defined route, a migration stream often flows from a highly restricted point of origin. The pool of potential migrants is kin-defined, often quite narrowly, because the informational links that convey data concerning optimal routes and destinations tend to follow kinship connections.

While we have no direct evidence of a pool of migrants defined by kinship journeying from Alaska to California, these travelers likely were mostly people who spoke the same dialect of Proto-Aleut-Utian, that is, Utian (9.3). Typically, in small-scale societies, people who speak the same dialect are often related to one another through kinship and marriage, some of course more closely than others.

9.6.1.4 Return Migration

Major migratory streams tend to develop counter streams of people returning to their place of origin (p. 904), especially when opportunities at the place of origin are perceived by migrants to be greater than those at the destination. While the Utian migration provides neither linguistic nor archaeological evidence of such a counter stream, we should be aware of possible evidence for it in future investigations of Utian migration since counter streaming is apparently a process of human migratory behavior.

Chapter 10

Contributions, Controversies, and Continuing Study

10.1 MAJOR CONTRIBUTION

The central contribution of this book is the recognition and compilation of comparative evidence for the genealogical relation of languages never previously deemed connected: Aleut of Alaska and Utian languages of California (chapters 1-4). An equally important contribution is the story of how speakers of these affiliated languages could have become so distantly removed geographically from one another (chapters 6 and 9). We propose migration by watercraft. To our knowledge, this is the earliest known, documented, ancient long-distance migration over water by Native Peoples of the Western Hemisphere.

10.2 ADDITIONAL CONTRIBUTIONS

10.2.1 Evaluation of Proposals of Language Relationship

We have described how two different approaches for comparative linguistics, quantitative and traditional, sometimes viewed as competitive, can be combined in a productive manner (chapters 1 and 2). The quantitative approach enables a systematic screening of comparative lexical data for word similarities likely due to genealogical relationships. It also incorporates an automated mechanism to evaluate lexical similarities, ASJP (chapters 1 and 9), which can also be used to date proto-language divergence (9.2). Applying quantitative screening yields numeric metrics for proposals of language relationships that are systematically connected with prosaic evaluations of supporting evidence, for example, *Strong*, *Moderate*, or *Weak for Kinship*. If evidence is evaluated as *Strong for Kinship,* then genealogical affiliation is considered verified for the proposal; if *Weak for Kinship*, then the proposal is unlikely and not further studied; if *Moderate for Kinship*, the proposal may or may not be valid and should be further examined to find out which one of these possibilities is more likely.

Applying our quantitative method to evaluate the strength of evidence for a link between Aleut and Utian yields the result *No Support for Connection*, indicating the proposal should be abandoned. Another, indirect approach is the application of our quantitative method to Aleut/Proto-Miwok, Proto-Miwok being an immediate daughter language of Proto-Utian. The result for Aleut/Proto-Miwok is *Moderate for Kinship* (chapter 2), suggesting further examination may be fruitful. Subsequent application of the traditional approach to Aleut/Proto-Miwok yields a positive result involving the reconstruction of sixty-six Aleut-Utian cognates, supported by twenty-six regular sound correspondences (chapter 2, 2.4.6).

One of the most critical issues in historical linguistics is distinguishing words with similarities due to kinship from those due to borrowing. None of the few guidelines yet proposed for doing this in a systematic manner have been more than moderately successful. Until a consistently effective approach is developed, any language-relatedness proposal for which extensive lexical resemblances have been compiled, especially those with only a few or no sound correspondences, can be legitimately challenged by critics on the grounds that similarities are due to borrowing rather than to language kinship.

Quantitative screening (chapter 1) involves an algorithm capable of sorting groups of words with similarities due to borrowing from those with resemblances due to inheritance (1.2.2). The algorithm distinguishes between basic vocabulary, words highly resistant to borrowing, and non-basic (or cultural) vocabulary, words not resistant to borrowing, and provides evaluative numerics such as those interpretable as *Strong* , *Moderate*, and *Weak for Borrowing*. If evidence for a proposal is found *Strong for Borrowing*, all lexical similarities observed are considered to be due to borrowing; if *Weak for Borrowing*, no supportive evidence for the proposal is observed; if *Moderate for Borrowing*, lexical similarities observed for a proposal may be due mainly to borrowing, and additional scrutiny may help to decide if this is likely. (Other evaluations used are *Strong for Connection, Moderate for Connection*, and *Weak for Connection*.)

We hope the compilation of results from the application of quantitative screening to evidence for 232 proposals of language genealogical relationships, provided in online appendix A, will be of use to comparative linguists. These include binary lexical comparisons of globally distributed languages, most of which are from published works as well as dissertations and theses.

10.2.2 Mitochondrial DNA Evidence for Aleut-Utian

Chapter 4 compiles evidence from human molecular genetics, specifically mitochondrial DNA (mtDNA), for the biological relationship of speakers of different languages ascertained by linguists to be genealogically related. Our focus is on speakers of Aleut and Utian languages, for which mtDNA shows a special genetic affiliation. In online appendix C, we present mtDNA haplogroups for 165 populations of Native Americans of North America, comprising both archaeological and contemporary samples gathered from the literature.

10.2.3 Ecological Setting

Chapter 7 describes ethnobiological settings likely encountered by speakers of Proto-Utian on their Middle Holocene migration along the Pacific coast of North America from southwest Alaska to central California. While scientific coverage of fauna is sparse, that of flora is considerable due to the abundance of identifiable pollen recoverable for the era in question. The chapter represents a Herculean effort in paleo-botanical reconstruction, with very few significant gaps in coverage of the approximately 5000 kilometers from Alaska to California, and shows how snapshots of the distant geo-biological past can be combined to provide a broad and easily interpretable picture of the natural settings in which Utians found themselves on their challenging journey.

10.2.4 Archaeological Perspective

Chapter 8 uses archaeology to supplement and enhance what we know about Proto-Aleut-Utian people acquired through linguistics and human molecular genetics. Among several investigative strategies, three archaeological models are employed: population replacement, immigration, and population integration. Each of these models entails unique expectations for archaeological findings relating to Utian migration. In examining archaeological evidence for the migration, we focus on those regions with the strongest linguistic evidence for visitation (central and southern Northwest Coast) and where Utians eventually settled (coastal central California). A major conclusion is that archaeological findings are most in accord with the population-integration model, whereby a small number of travelers move into a new area and create peaceful relationships with the resident population, resulting in the bi-directional transmission of language, culture, and genes.

Migration of Utians to California apparently did not involve conflict. Rather, Utian migrants interacted closely with extant residents when colonizing the San Francisco Bay Area in the Middle Holocene. Shared with resident groups, Utians built upon a maritime focus on shellfish, fish, sea mammals, and on a local cultural repertoire involving acorn and small seed harvesting. This led to the creation of the imposing shell mounds that eventually populated the entire greater San Francisco Bay Area. This model of cultural integration is consistent with Moratto's (1984:553) idea for the fusion of extant Native California populations with Utian newcomers. Thus, the ancestry of historical Costanoan-Miwok people likely includes not only an ancient lineage from Alaska, but also endemic California groups who have resided there since at least the Late Pleistocene and Early Holocene.

10.3 CONTROVERSIES

10.3.1 California Penutian

Our discussion of California Penutian is found in chapter 5 (5.3). The California Penutian proposal unites Utian (Costanoan and Miwok), Maidun, Wintun, and Yokuts in a single genealogical group. This proposal is now over one hundred years old and has remained controversial for most of its history. Scholars have been engaged in arguments as to whether the extensive similarities, primarily involving the lexicon, observed for California Penutian are due to borrowing or to language kinship.

In chapter 5, we apply quantitative screening to various published lexical comparisons of California Penutian languages, with varying results. As expected, evidence for a Costanoan-Miwok connection is *Strong for Kinship*. This connection defines the Utian group of languages, for which we find substantial evidence for a genealogical linkage with Aleut, and is the only binary California Penutian linkage whose evidence for kinship is determined *Strong*.

For other California Penutian binary comparisons, evidence is *Moderate for Kinship* for Utian-Wintun and Maidun-Yokuts connections; *Moderate for Connection* for both Utian-Yokuts and for Wintun-Maidun; and *Moderate for Borrowing* for Utian-Maidun (figure 5.1). All the mentioned connections may be either verified or rejected through further research. At this time, the best graphic presentation of California Penutian language affiliations we can provide is the language-relationship map in figure 5.1. Unfortunately, this map does not bring us to a conclusive result in sorting out the California Penutian tangle of language associations, but it is a useful start.

10.3.2 Eskimo-Aleut

Only very recently has the long-held (around 200 years) proposal of a genealogical affiliation for Proto-Eskimo and Aleut been called into question.

Credit for this goes to Anna Berge (2018) who, in her work challenging the status of many similar words of the two languages long thought to be cognate, suggests instead that these are loanwords. In chapter 5 (5.5), we apply quantitative screening to lexical evidence for Eskimo-Aleut, yielding an evaluation of *Moderate for Borrowing*. This result indicates that any historical connection between Eskimo and Aleut is through contact and borrowing rather than kinship. The amount of vocabulary shared by Eskimo and Aleut is very large, indicating borrowing between the two languages must have occurred over an extended period of time (9.3).

In chapter 9, we assemble evidence for homeland locations for both Proto-Eskimo (9.4.3) and Early Aleut (9.4.4). These proposed locations are close in proximity to each other, with the Proto-Eskimo homeland in the northeast of the Alaska Peninsula and the Early Aleut homeland next door in the old Proto-Aleut-Utian region of greater Kodiak Island. This proximity would have made speakers of Aleut and Proto-Eskimo very accessible to one another, thus facilitating language borrowing. Moreover, both groups may have occupied these homeland areas for long periods of time, perhaps thousands of years (see figure 9.1).

As noted in chapter 4 (4.4), mtDNA haplogroup profiles for Eskimo and Aleut populations indicate some similarity for these two groups (comparison score 2), which might, we suppose, accord with close and prolonged contact.

10.3.3 Dene-Yeniseian

The results of our quantitative investigation of a proposed relationship between the Yeniseian language family of Central Siberia and the Na-Dene family of North America (Alaska, California, Canada, Arizona, and New Mexico) are presented in online appendix A (see comparison 182). This is among the most controversial proposals of contemporary historical linguistics. Evidence for the hypothesis is assembled by Vajda (2010) in a chapter of a volume edited by Kari and Potter (2010). Most of the volume's papers treat aspects of Vajda's proposal, including articles by comparative linguists, archaeologists, human geneticists, folklorists, and others. These are about equally "thumbs up" and "thumbs down" for the hypothesis, but most urge caution. (Vajda with Fortescue has recently published a book [Fortescue and Vajda 2022] in which Vajda discusses his proposal and related developments since the 2010 work [see below].)

We apply quantitative screening (chapter 1) to sets of presumed cognates for Yeniseian and Na-Dene languages presented in Vajda (2010). The result is *No Support for Connection*, indicating that further linguistic investigation (specifically, use of the traditional comparative method) is not likely to yield a positive report. This result does not preclude looking at other kinds of evidence, however, such as that from archaeology and human molecular genetics, and such interdisciplinary approaches have been applied to the Dene-Yeniseian hypothesis in the Kari and Potter (2010) book; see Potter (2010:138–167, for archaeology) and Scott and O'Rourke (2010:119–137, for human genetics). Potter's (p. 159) main conclusion is:

> Given the patterns of archaeological variability in this region, I believe the archaeological record alone will not provide certainties regarding correlation of language, genetic populations, and material culture, but may offer a framework to evaluate cognate sets and elucidate time depth of language divergence.

Regarding human genetics, Scott and O'Rourke (p. 133) write: "While linguists have developed a solid case linking the North American language family Na-Dene with Yeniseian, we have not found comparable parallels in the biology of these 'groups'."

After the passage of more than a decade, in which new archaeological and genetic data have emerged, in his 2022 book with Fortescue, Vajda's own evaluation of the situation is: "human genetic and archaeological links between Yeniseian and Na-Dene speakers—whether direct or mediated through admixture with different peoples—cannot determine whether the languages ancestrally spoken by the two groups were in fact related."

Given these assessments and our negative result using quantitative screening, it now seems appropriate to outrightly reject the Dene-Yeniseian hypothesis. But at least one more avenue of investigation is open before this extreme measure is taken: the quantitative evaluation of the sharing of structural principles (chapter 3) by Yeniseian and Na-Dene languages to determine whether the extent of sharing of these features surpasses chance expectation. If so, a historical connection between Dene-Yeniseian and Na-Dene is likely, either through language contact (promoting borrowing) or genealogical affiliation (language kinship).

We undertake this evaluation using the online version of the *World Atlas of Language Structures* (Dryer and Haspelmath 2013, http://wals.info/) or simply *WALS*. Since we use *WALS* data extensively in chapter 3, we do not repeat here the detailed description of the online source or our explanation of how it is used.

For this investigation, we compare structural features and their values in Yeniseian and Na-Dene languages found in *WALS*. For example, Ket, a Yeniseian language, has 124 entries (language features), two examples of which are 13A Tone and 68A The Perfect. Each of these 124 features has two or more values. For 13A, there are three values: "No Tones," "Simple Tone System," and "Complex Tone System" (Ket is "No Tones"); for 68A, there are five: "From Possessive," "From Finish," "Already," "From Other Perfect," and "No Perfect" (Ket is "No Perfect").

Our evaluation uses this information to ascertain if Ket and Na-Dene languages share values beyond what is expected by chance. If so, we can confidently conclude that a historical connection for Ket/Na-Dene languages exists, but not, unfortunately, the specific nature of the language connection, that is, either contact or kinship. (Only Ket, of the four *WALS* Yeniseian languages, is used since very little information is reported for the other three, all of which are extinct.) Online appendix F reports the results of our evaluation. Findings exceed chance expectation, unexpectedly indicating a connection between Ket and Na-Dene languages.

10.4 CONTINUING STUDY

10.4.1 Lines of Evidence

Five lines of evidence for the Aleut-Utian(-Miwok) proposal are used in this book: quantitative screening (chapters 1 and 2), the traditional comparative method (chapter 2), structural features (chapter 3), mitochondrial DNA (chapter 4), and archaeology (chapter 8). Table 10.1 presents support for each of three language-relatedness proposals provided by these lines of evidence. The three proposals are: Aleut-Utian(-Miwok), the featured hypothesis of this book; Dene-Yeniseian, described in 10.3.3; and Athapaskan, the long-held non-controversial proposal of the relationship of Subarctic and U.S. Southwest Athapaskan languages (see 10.4.2 for more information on Athapaskan).

In table 10.1, evidence from the traditional comparative method, structural features, and mtDNA are all rated *Supportive* for Aleut-Utian(-Miwok); that

Table 10.1 Five Lines of Evidence for Three Language-relationship Proposals

Line of Evidence	*Aleut-Utian(-Miwok) (Middle Holocene)*	*Dene-Yeniseian (6700–5900 BP*)*	*Athapaskan (2500–2200 BP*)*
Structural features	Supportive	Supportive	Supportive[1]
Comparative method	Supportive	Not supportive	Supportive[2]
Quantitative screening	Moderately supportive	Not supportive	Supportive[2]
Mitochondrial DNA	Supportive	Not supportive	Supportive
Archaeology	Not supportive	Not supportive	Supportive

Source: Generated by the authors.
*These dates are from Fortescue and Vajda (2022:169).
[1]59.0 = SVP (structural value percentage [chapter 3] for comparison of Navajo [Southern Athapaskan] and Slave [Canada Athapaskan]).
[2]These results are assumed (rather than observed), based on knowledge of the shallow chronological depth of the Athapaskan family, compared to the great chronological depths of Aleut-Utian and Dene-Yeniseian groups.

of quantitative screening is *Moderately Supportive* for Aleut-Utian(-Miwok), indicating further study may be fruitful (as it has been), and evidence from archaeology is *Not Supportive* in the sense that uniquely similar artifacts from Early Aleut and Proto-Utian sites respectively are ambiguously attested if found at all.

For Dene-Yeniseian (table 10.1), four of the five lines of evidence are not supportive of a connection between Na-Dene and Yeniseian (10.3.3). The proposal *is supported*, however, by the similarity of structural features, defying the expectation of chance occurrence (online appendix F). Since the associated statistics are highly statistically significant, we embrace the likelihood that Na-Dene and Yeniseian are historically connected (10.3.3.1). Nevertheless, we leave open the question of whether these languages are genealogically related since a historical connection could entail either language contact or language kinship. A search for similarities involving structural features beyond those of the *WALS* database is warranted.

Table 10.1, evaluations for Athapaskan (all five lines of evidence are supportive) come in part from a genetics paper (Malhi et al. 2008) looking at Y chromosome distribution among Native North Americans. Despite some comments seeming to suggest otherwise, these authors recognize linguistic, genetic, and archaeological evidence coalescing in support of a relatively recent migration of a small group of Subarctic people to the U.S. Southwest, conforming with the Athapaskan migration hypothesis. (See also populations 34–36, and 63–67 of online appendix C.)

Among the five lines of evidence, only "structural features" are ranked *Supportive* for each of the three proposals. Given this finding, comparative linguists should perhaps give initial attention to the possible sharing of structural features by compared languages before using other kinds of evidence to explore relationships, especially those of great temporal distance. Furthermore, only Athapaskan shows the same across-the-board *Supportive* evaluation for all five lines of evidence. No other consistency in results is apparent across lines of evidence reported in table 10.1. This accords with a quote of Boas made over one hundred years ago, which seems just as valid today as it did over a century ago:

> These considerations make it fairly clear that, at least at the present time, anatomical type [*our genetics*], language [*our linguistics*], and culture [*our archaeology*] have not necessarily the same fates, that a people may remain constant in type and language and change in culture, that they may remain constant in type, but change in language, or that they may remain constant in language and change in type and culture. (Boas 1911:11) (bracketed comments are supplied by authors.)

10.4.2 Grammatical Morphemes

Table 10.1 may be considered incomplete by some since it does not include grammatical morphemes as another line of evidence for language relatedness. Grammatical morphemes are phonological elements marking values of grammatical categories such as past, present, and future for tense; singular, plural, and dual for number; male, female, and neutral for gender; and so on. Comparing grammatical morphemes across languages for cognation is viewed by some comparative linguists as an essential, if not the most important, component of comparative study (cf. Fortescue and Vajda 2022).

Grammatical morphemes should not be confused with structural features of language (chapter 3), even though the latter are often grammatical. For example, two languages can be compared for whether they show duality of number (a structural comparison) or, if they show duality and have morphemes denoting such, whether those morphemes are similar or, perhaps, cognate (a grammatical morpheme comparison).

It is difficult to understand how morphological evidence might somehow be more probative of genealogical relationship than lexical evidence. Reasons for not viewing grammatical morphemes as a major line of comparative evidence are compelling. Since morphological particles are typically less complex (shorter) than other (referential) words, they are considerably more likely to be similar by chance when compared, thereby promoting the development of coincidental similarity across languages. Campbell (1998:322) writes, "Monosyllabic words (composed of a single consonant and vowel) may be true cognates, but they are also so short that their similarity to forms in other languages could also easily be due to chance."

A list of all grammatical morphemes (15 in number) reconstructed by Vajda (Fortescue and Vajda 2022) for Proto-Dene-Yeniseian follows. These are taken from a compilation of 140 lexical reconstructions for the language, retaining their original ID numbers:

10) ~*de- [valence lowering prefix]
33) ~*-tɬ "with"
38) ~*-ɬ- [imperfective affix]
46) ~*si- [conjugation prefix that alternates with conjugation prefix *ɢa-]
47) ~*s- [action nominal prefix, usually called the "gerund" in Na-Dene scholarship]
99) ~*ɢa- [conjugation prefix, alternating with *si-]
100) ~*-qej [locational suffix]
130) ~*wu [third person animate pronoun]
131) ~*w- [third person verb agreement prefix, probably etymologically connected with the preceding morpheme]

132) *j- [third person verb agreement prefix]
136) ~*-ŋʷ - [change-of-state affix]
137) ~*-ji [resultative affix]
138) ~*-ŋʷ [action nominal / gerund suffix, appears together with action nominal / gerund prefix *s-]
139) ~*-ŋʷ - [possessive connector]
140) ~*-ej [kin noun suffix]

Only one grammatical morpheme of the fifteen shows more than two sound segments (see no. 100). Among the other fourteen, seven show only one segment, and seven show just two segments.

The literature provides considerable evidence for the lack of resistance of grammatical morphemes to borrowing. A few examples include Aikhenvald 1996, Beck 2000, 2023, Comrie 1989, Epps and Bolaños 2017:491–495, Heath 1978, Nichols 1992, and the cross-language treatment of this and related topics in Thomason and Kaufman 1988. In this respect, grammatical morphemes are more like words denoting cultural items (not resistant to borrowing) than words of basic vocabulary (resistant to borrowing).

10.4.3 Ships Ahoy! More Ancient Mariners

10.4.3.1 Athapaskan Ancient Mariners

Examining the evidence assembled below, we hypothesize that, roughly 3000 years after the marine migration of Utian people from southwest Alaska to the San Francisco Bay Area, prehistoric peoples speaking an Athapaskan language departed on watercraft from approximately the same locale in southwest Alaska, destined for coastal areas of north California and southwest Oregon, where they disembarked and permanently settled.

With a shallow ASJP date range of 2660–1464 BP (Holman et al. 2011), Athapaskan is a large family of languages spoken in western North America that includes three discontinuous geographic groupings. The largest, Northern Athapaskan, extends from Alaska, the Yukon, and Northwest Territories into parts of British Columbia (BC) and Alberta. The two others are both considerably smaller in geographic spread (and number of languages): the Apachean languages of the U.S. Southwest (including Navajo or Diné, and Apache), and Pacific Coast Athapaskan of north California and south Oregon.

Pacific Coast Athapaskan languages for which relevant linguistic information is available include four in California (Hupa, Wailaki, Mattole, and Kato) and two in Oregon (Galice and Tututni), with one more language, Tolowa, straddling the state line. According to Hoijer (1963:9), these languages form a genetic subgroup within Athapaskan, with two divisions, California and Oregon, although these distinctions today seem yet to be settled (see especially Snoek 2015).

As far as we can determine, only Tolowa speakers among the Pacific coast groups now inhabit areas directly adjoining ocean waters. Moreover, only these people harvest marine mammals such as sea lions (Gould 1978:130). Apparently, the remaining Pacific Coast Athapaskans, while living relatively close to the sea, are basically interior—rather than marine-oriented, a pattern shared with all Northern Athapaskan groups save one, speakers of Dena'ina (sometimes written Tanaina), whose coastal populations adopted a maritime-harvesting subsistence economy (Workman 1998:146).

Dena'ina is spoken in southwest Alaska around Cook Inlet and inland therefrom (Kari 2020). This language seems to show no special affiliation with any other Athapaskan language and is unusual among Northern Athapaskan languages since its speakers alone within this language family occupy any substantial area of coastline. The historic region of Dena'ina speakers is part of the coastal region of the Gulf of Alaska we identify in chapters 8 and 9 as the Proto-Aleut-Utian homeland, that is, areas of the Alaska Peninsula, Kodiak Island, and the Kenai Peninsula, populated today by speakers of Alutiiq Yupik Eskimo as well as Dena'ina speakers (figure 10.1).

The geographic disjunct between Pacific coast and Northern Athapaskan groups implies an ancient event (or events) in which the ancestors of these respective peoples became separated through migration. Determining the direction of movement depends in part on where ancestors of all Athapaskan

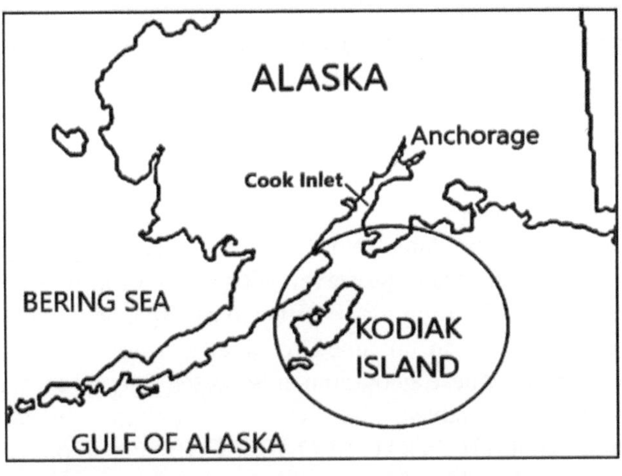

Figure 10.1 Kodiak Archipelago and Immediately Abutting Areas of Coastal Southwest Alaska. *Note*: The encircled area is the approximate location of the proposed Proto-Aleut-Utian homeland (9.4.1). *Source*: Generated by the authors.

speakers, that is, the Proto-Athapaskans, were located. The computer algorithm of Wichmann et al. (2010) supports a BC location as the most likely PA homeland. Linguists also look to Northern Athapaskan for a likely site, implying that ancestors of Pacific Coast Athapaskans were the migrant group who moved from north to south, not the reverse. Krauss and Golla (1981:68) narrow the PA location even further to "eastern interior Alaska, the upper drainage of the Yukon River, and northern British Columbia or some part of this area."

Krauss and Golla (1981:69) also propose from where Pacific Coast Athapaskans might have begun their migration to California/Oregon. They write, "these languages appear to have been an offshoot from the British Columbia [Athapaskan] languages," although they provide no firm evidence for this proposal. They also fail to mention the BC location from which the migration may have begun or the route it may have taken. One possibility is that the journey, like the migration of Proto-Aleut-Utian people from Alaska to California thousands of years earlier, could have been by watercraft across the ocean.

Speakers of the now extinct Ts'ets'aut language, located at the northern end of Portland Inlet in the far northwest corner of British Columbia, settled closest to the Pacific Ocean coast among all historic BC Athapaskan groups. Nevertheless, these people were removed by a considerable distance from the coast through a complex of long natural outlets. Furthermore, we have not been able to determine any special linguistic affiliation for Pacific Coast Athapaskan with the Ts'ets'aut language or, for that matter, with any other Athapaskan language of BC.

Our linguistic investigation uses a very large data set, provided in part by Conor Snoek in his dissertation "The Lexical Semantics of Athapaskan Anatomical Terms: A Historical-Comparative Study" (2015). The lexical items included in Snoek's database are taken exclusively from a single semantic domain, the human body and its by-products, including seventy-two anatomical "referent-concepts" lexically encoded by 2692 words in fifty-five mostly well-documented Athapaskan languages. Data are organized so that every available anatomical word of any Athapaskan language is easily compared with every available anatomical word of other Athapaskan languages. The fifty-five languages are sorted into seven geographic groups: Alaska, Yukon, Interior Canada, British Columbia, Pacific Coast, Apachean (U.S. Southwest and abutting areas), and an unnamed residual area for only one language, Tsuut'ina of Alberta. These geographically sorted languages and terms are presented in sixty-eight tables, each table dealing with a different referent-concept such as MOUTH, FOOT, BLOOD, SCAB, and so on. (Our analysis excludes Tsuut'ina of Alberta.)

We manually search Snoek's (2015) large database to find body-part words exclusive in occurrence with one or more of the Pacific Coast Athapaskan languages and with one or more languages of the other geographic groups mentioned above. We assume the more frequently an Athapaskan language of a specific region is observed to have lexical items for body parts closely phonologically resembling those of Pacific coast languages, the greater the likelihood that speakers of the former languages were involved in a special connection with Athapaskan people of the Pacific coast.

The results of our search of Snoek's anatomical vocabulary are reported in table 10.2. Four body-part terms (ABDOMEN, BODY, FINGER, and HEART) are exclusive to Pacific coast languages and languages of one other Athapaskan geographic group, in all instances Alaska. We also searched our own extensive collections of words in Athapaskan languages for plants and animals, finding terms for ALDER, WILLOW, and WHALE limited in occurrence to Pacific coast languages and one other geographic group, again, all Alaska. From these results, we conclude that some Athapaskan language or languages of Alaska have a special connection with the Pacific Coast Athapaskan languages, probably involving migration.

Table 10.2 Words for Body Parts, Plants, and Animals Exclusive to Pacific Coast Athapaskan Languages and Just One Other of Six Geographic Regions in Which Athapaskan Languages Are Spoken, Based Mainly on Snoek (2015)

	Abdomen	*Alder*	*Body*	*Finger*	*Heart*	*Whale*	*Willow*
Alaska							
Dena'ina	*kʰun*	*q'esh*	*nest'a*	*luʒəh*	*kʰuzʔin, kʰujʔin*	*tałin*	*q'eylish*
Ahtna	*kʰan*					*tełaani*	
Deg Xinag	*ʧʰon̥*						
Koyukon	*kʰɒn̥*						
Pacific Coast							
Hupa	*kʲa:n*	*q'iwh*[1]	*nɪst'*		*kʲʰʌnsaʔa:n*	*tehla:n*	*q'ayliwh*[1]
Galice			*tast'*				
Kato		*k'ash*	*nəst'*	*la:sak'*		*teehlaang*	
Tolowa		*g'ush*				*tee-lan'*	*ge'-liish*
Tututni				*lasək'*			
Apachean							
Navajo		*k'ish*[2]					
Yukon							
Interior Canada							
British Columbia							

Sources: Table organized by the authors. All body-part words are from Snoek (2015). *Sources* for other words in the table are Ahtna (Kari 1990), Dena'ina (Inland) (Tenenbaum 1975), Dena'ina (Outer Inlet [Kenai]) (Kari 1977; see also Kari, P. R. 2020), Hupa (Golla 1996), Kato (Anderson 2018), and Tolowa (Boommelyn 2006).

[1]Stem-final *-sh* from Dena'ina in both examples remains *-sh* in Tolowa and Cahto but shifts to *-wh* in Hupa.

[2]This Navajo (Apachean) word for "alder" is unexpected given its occurrence in Athapaskan languages of both Alaska and the Pacific coast. We have no explanation for this at present.

In each of the seven sets with words found exclusive (one semi-exclusive) to languages of the Pacific coast and Alaska, Dena'ina is the only Alaskan language common to all. Dena'ina would then seem a likely candidate for a single-language source for the Pacific coast terms. Furthermore, in all but one set, that for FINGER, Hupa is one of the Pacific coast languages, suggesting that Hupa may have been the original Pacific coast language to have the words in question. Could Dena'ina and Hupa have been dialects of the same language diverging from one another after Hupa speakers migrated from south Alaska, leaving Dena'ina speakers behind?

How did Athapaskan peoples transport themselves from southwest Alaska to coastal California/Oregon (approximately 2600 km as the crow flies)? Since speakers of PAU languages migrated by watercraft some 3000 years earlier (chapters 6, 8, and 9), envisioning a similar seafaring journey by a much later prehistoric group is not difficult, considering advances in watercraft technology and navigation skills likely to have developed during the several millennia after PAU speakers undertook their migration to California.

The launching of both earlier and later prehistoric migrations from the same port in coastal southwest Alaska, in or near to the PAU homeland (the greater Kodiak region), may not necessarily have been coincidental. The natural features of this coastal area, especially many small, protective bays, would support the development and use of watercraft for both hunting marine creatures and transport in the location, and this would have developed as common knowledge among both coastal and non-coastal peoples of south Alaska in prehistoric times. Even today, Cook Inlet provides the only water access from the Gulf of Alaska to Anchorage (figure 10.1), Alaska's most populous city and largest seaport.

In 10.4.1, we evaluate five lines of evidence possibly supportive of proposals of language affiliation: quantitative screening (combined with ASJP), the traditional comparative approach, structural features, human molecular genetics, and archaeology. We also discuss a sixth line, grammatical morphemes (10.4.2), but dismiss this evidence of possible genealogical connection since the development of similarity is likely strongly chance influenced.

One other source of evidence has been mentioned, but only briefly, several times in chapters 7 and 8 (e.g., 7.4, 8.3.1). This evidence comprises oral histories collected from speakers of target languages. Possibly due to the great geographical distance and chronological depth of the connection between Aleut of Alaska and Utian of California, any direct oral evidence pertaining to this central focus of our research has yet to emerge. An important folk narrative collected and reported by David Lewis (2023) does relate, however, to the proposed Alaska Athapaskan/Pacific Coast Athapaskan connection. He writes, "According to Tolowa oral histories, the Athapaskan people of southern Oregon and northern California arrived from the north in ancient times, traveling by canoe."

10.4.3.2 Other Ancient Mariners: Chumash?

Chumashan is a group of several closely related languages of southern coastal California (see maps of figures 6.1 [Ineseño] and 10.2). None of these languages has yet to be successfully genealogically linked to any other language or languages outside their group. For this reason, Chumashan speakers are thought by many to be among the oldest Native Residents of California. In this final section, we examine mtDNA evidence for modern Chumash people, comparing it to that of other western North American populations. This evidence indicates similarities of Chumash to several other populations, suggesting ancestors of modern Chumash migrated to California using watercraft and thus should be included with the two other groups of possible ancient mariners identified in this book, Utians and Athapaskans.

The database for our discussion is presented in table 10.3. This information is extracted from online appendix C, as is the information of table 4.1. In both of these sources, data are displayed in identical ways, and this common means of presentation is described in sections 4.2 and 4.3.

Chumash is population ID 25 in table 10.3. Haplogroup profiles of the remaining twenty-seven groups of table 10.3 are all similar to that of Chumash since all have both most-frequent *A* and second-most-frequent *D* lineages. Of the twenty-seven groups, five show *exactly* the same profile as that observed for Chumash (comparison scores 4). These five are given in table 10.4 with their ID from table 10.3, with family affiliation, and with distance in kilometers from Chumash (See section 4.2 of chapter 4 for an explanation of gray shading in presentation of haplogroup profiles.)

All five of these populations are closely *genetically* related to the Chumash group and closely to one another (all possible binary comparisons for these six languages show a comparison score of 4). Also, since Chumashan is a language isolate, none of the five languages is assumed to be *genealogically* linked to it. Some of the five populations are closer *geographically* to Chumash than others are. The closest to Chumash (Obispeño) is Nuu-chah-nulth, (Wakashan) of British Columbia, removed from Chumash (Obispeño) by 1670 km; the most distant are Siberian Yupik (Eskimo) and Gambell Eskimo by 4580 km (see table 10.4 for other distances). Also, all five language populations are *coastally* located, as is Chumash (see map in figure 10.2).

The geographic distribution of the Chumash and the five groups of table 10.4 across western North America is almost certainly the result of population migration. Presumably, a single population having the Chumash haplogroup profile began a migratory spread, resulting in the distribution observed for the five language populations of table 10.4 plus Chumash. We can only guess

Table 10.3 Twenty-Eight Populations All with Both Most-Frequent *A* and Second-Most Frequent-*D* Haplogroups Extracted from the 165 Populations of Online Appendix C

Region 1: Modern Arctic									
ID	*Language*	*Family*	*Location*	N	*A*	*B*	*C*	*D*	*Key*
1	Siberian Yupik 1	Eskimo	East Siberia	79	77	0	3	20	1
2	Ouzinkie Eskimo	Eskimo	Alaska, Kodiak Island	41	73	0	5	15	2
3	Gambell Eskimo	Eskimo	Alaska	50	58	0	14	26	2
4	Old Harbor Eskimo	Eskimo	Alaska, Kodiak Island	115	62	3.5	0	35	2
5	Siberian Yupik 2	Eskimo	East Siberia	50	80	0	0	20	3
6	Sireniki	Eskimo	East Siberia	37	70	0	0	30	4
7	Siberian Eskimo 1	Eskimo	East Siberia	50	90	0	0	10	4
8	Naukan	Eskimo	East Siberia	39	47	0	0	26	4
9	Canadian Inuit	Eskimo	Canada	96	88	0	0	13	4
10	Greenland Inuit	Eskimo	Canada	385	96	0	0	4	4
11	Savoonga Eskimo	Eskimo	Alaska	49	94	0	0	2	2
12	Inuit	Eskimo	Alaska and Canada	30	97	0	0	3.3	5
13	Aleut (King Cove)	Aleut	Aleutians and Bering	33	52	0	0	49	6
14	Aleut (Nelson Lagoon)	Aleut	Aleutians and Bering	16	56	0	0	44	6
15	Aleut (Unalaska)	Aleut	Aleutians and Bering	28	61	0	0	39	6
Region 1: Ancient Arctic									
ID	*Site Name*	*Date*	*Location*	N	*A*	*B*	*C*	*D*	*Key*
16	Sadlermiut	977–682 BP	Canada, Hudson Bay	18	56	0	0	44	7

Region 2: Modern Subarctic									
ID	*Language*	*Family*	*Location*	N	*A*	*B*	*C*	*D*	*Key*
17	Chipewyan	Athabaskan	Canada	40	100	0	0	0	8
18	Dogrib2	Athapaskan	Canada	42	100	0	0	0	5
19	Bella Coola 1	Salish	British Columbia	25	60	8	8	20	2
20	Haida1	Haida	Canada and Alaska	25	96.6	0	0	3.4	3
21	Nuu-Chah-Nulth1	Wakashan	British Columbia	102	45.1	6.9	15.7	25.5	9
22	Bella Coola 2	Salish	British Columbia	36	50	5.6	14	25	5
23	Laxgalts'ap	Tsimshianic	British Columbia	20	95	0	0	0	8
REGION 3: Modern Northwest Coast									
ID	*Site Name*	*Date*	*Location*	N	*A*	*B*	*C*	*D*	*Key*
24	Bella Coola	500–1500 BP	British Columbia	84	65.5	10.7	9.5	14.3	9
REGION 4: Modern California (Outside of San Francisco Bay Area)									
ID	*Language*	*Family*	*Location*	N	*A*	*B*	*C*	*D*	*Key*
25	Chumash	Isolate	N. Channel Islands	21	52.4	0	9.5	38.1	10
Region 9 Modern Southeast									
ID	*Language*	*Family*	*Location (historical)*	N	*A*	*B*	*C*	*D*	*Key*
26	Creek/Seminole	Musgokean	Alabama, Georgia	18	38.9	11.1	16.7	16.7	5
Region 11: Modern Middle America									
ID	*Language*	*Family*	*Location*	N	*A*	*B*	*C*	*D*	*Key*
27	Kuna	Chibchan	Panama, Colombia	16	100	0	0	0	2
28	Nahua-Coyolillo	Uto-Aztecan	Mexico	38	68.4	7.9	0	15.8	11

Sources: Compiled by authors from this data: 1. Schurr et al. 1999, 2. Merriwether et al. 1995, 3. Torroni et al. 1993, 4. Volodko et al. 2008, 5. Lorenz and Smith 1996, 6. Crawford et al. 2010, 7. Raff et al. 2011, 8. Hughes et al. 2016, 9. Malhi et al. 2004, 10. Johnson and Lorenz 2006, 11. Peñaloza-Espinosa et al. 2007.

Table 10.4 Five Language Populations of Table 10.3 Having the Same Haplogroup Profile as Chumash with Distances (km) Removed from Chumash

ID	*Language Population*	*Language Family*	*Location*	*Km from Chumash*
1	Siberian Yupik 1	Eskimo	St. Lawrence Island, Alaska	4,580.0
2	Ouzinkie Eskimo	Eskimo	Kodiak Island, Alaska	3,463.0
3	Gambell Eskimo	Eskimo	St. Lawrence Island, Alaska	4,580.0
21	Nuu-Chah-Nulth 1	Wakashan	British Columbia, Canada	1,670.0
22	Bella Coola 2 (Nuxalk)	Salish	British Columbia, Canada	1,954.0

Source: Generated by the authors.

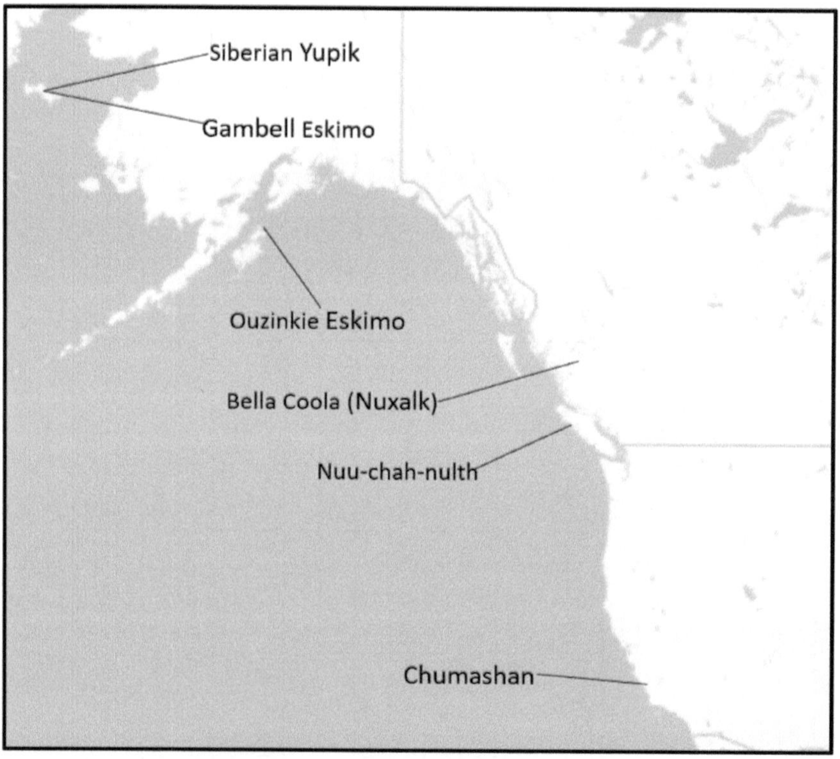

Figure 10.2 Map of Coastal Western North America with Location of Populations Having the Chumashan mtDNA Haplogroup Profile. *Source*: Map outline adapted from Dryer and Haspelmath (2013). Permission granted by Martin Haspelmath.

where the original population was located, but since the movement of people from far East Asia into the Americas and beyond has generally been from north to south, so too may have been the direction of advancing people with the Chumash profile. The distribution of the five populations plus Chumash suggests a migratory chain beginning in west Alaska (St. Lawrence Island), to south Alaska (Kodiak Island), to British Columbia (Canada), and ultimately to southern California (see map, figure 10.2).

All five populations plus Chumash are coastally located. Siberian Yupik, Ouzinkie Eskimo, Gambell Eskimo, Nuu-Chah-Nulth, and Chumash are directly on the coast, and Bella Coola (Nuxalk) is on channels permitting easy access to sea waters. Such coastal locations suggest migration was facilitated through the use of watercraft, the journey possibly beginning in Bering Sea ports and terminating on the shore of southern California, with prolonged stops along the way in coastal southwest Alaska and coastal British Columbia.

While the mtDNA profile of these ancient mariners remained unchanged during migration, the language of these travelers (whatever its identification) would have been modified and perhaps mostly replaced by the languages of different populations they encountered along the way and to whom they donated their haplogroup profile. For example, the Chumashan language was spoken approximately in its present location before a group with the modern Chumash mtDNA profile arrived on the scene. Sometime after their arrival, these immigrants adopted the Chumashan language, and their mtDNA somehow became standard among all Chumashan-speaking people who survived to historic times.

Coda

This study is suggestive of the central role human migration has played, over water as well as land, for thousands, if not tens of thousands, of years. Any concerted effort by governments or individuals to impede or stop it is ultimately doomed to failure, regardless of the kinds of barriers chosen, physical or political.

References

Adams, William Y., Dennis P. Van Gerven, and Richard S. Levy. 1978. "The retreat from migrationism." *Annual Review of Anthropology* 7: 483–532.

Aikhenvald, Alexandra Y. 1996. "Areal diffusion in Northwest Amazonia: The case of Tariana." *Anthropological Linguistics* 38: 73–116.

Alutiiq Museum and Archaeological Repository. 2022. https://alutiiqmuseum.org/images/stories/collections/ATI/ATI_CH8c.pdf

Ames, Christopher J. H., Andre Costopoulos, and Colin D. Wren. 2010. "8,000 years of technological change in the Gulf of Georgia: Is there a major transition at 4850 cal B.P.?" *Canadian Journal of Archaeology* 34: 32–63.

Anderson, Gregory D. S., and K. David Harrison. *Siletz Talking Dictionary*. Living Tongues Institute for Endangered Languages. http://siletz.talkingdictionary.org. 2007.

Anderson, M. K. *Tending the Wild: Native American Knowledge and Management of California's Natural Resources*. Berkeley, CA: University of California Press. 2005.

Anderson, Sally R. *Cahto Dictionary*. SIL International. Webonary.org. 2018.

Anichtchenko, Evguenia. 2012. "Open skin boats of the Aleutians, Kodiak Island, and Prince William Sound." *Études/Inuit/Studies* 36(1): 157–181.

Anichtchenko, Evguenia. 2013. "Did Indigenous Arctic Mariners use sails before the European contact?" Presentation delivered in Trondheim, Norway on October 2, 2013, Smithsonian Arctic Studies Center, Southampton University, Southampton.

Anthony, David W. 1990. "Migration in archaeology: The baby and the bathwater." *American Anthropologist* 92: 895–914.

Applegate, Richard. 1992. "Chumash (Ineseño)." Unpublished manuscript, Santa Barbara Museum of Natural History, Santa Barbara, CA.

Atleo, E. Richard (Chief Umeek). *Principles of Tsawalk: An Indigenous Approach to Global Crisis*. Vancouver: UBC Press. 2011.

Atwater, B. F., S. G. Conard, J. N. Dowden, C. W. Hedel, R. L. MacDonald, and W. Savage. 1979. "History, landforms, and vegetation of the estuary's tidal marshes."

In *San Francisco Bay: The Urbanized Estuary*, edited by T. J. Conomos, pp. 347–381. San Francisco, CA: American Association for the Advancement of Science, Pacific Division.

Atwater, B. F., E. J. Helley, and C. W. Hedel. 1977. "Late Quaternary depositional history, Holocene sea-level changes, and vertical crustal movement, Southern San Francisco Bay, California." *United States Geological Survey Professional Papers* 1014: 1–15.

Aulestia, Gorka and Linda White. *Basque-English English-Basque Dictionary*. Reno, NV: University of Nevada. 1992.

Baichtal, J. F., and R. L. Carlson. 2010. "Development of a model to predict the location of early Holocene habitation sites along the western coast of Prince of Wales Island and the outer islands, Southeast Alaska." *Current Research in the Pleistocene* 27: 64–66.

Baichtal, J. F., A. J. Lesnek, R. L. Carlson, N. S. Schmuck, J. L. Smith, D. J. Landwehr, and J. P. Briner. 2021. "Late pleistocene and early holocene sea-level history and glacial retreat interpreted from shell-bearing marine deposits of Southeastern Alaska, USA*:"* *Geosphere* (2021) 17 (6): 1590–1615.

Barbieri, Chiara, Damian E. Blasi, Epifanía Arango-Isazaa, Alexandros G. Sotiropoulos, Harald Hammarstrom, Søren Wichmann, Simon J. Greenhill, Russell D. Gray, Robert Forkel, Balthasar Bickel, and Kentaro K. Shimizua. 2022. "A global analysis of matches and mismatches between human genetic and linguistic histories." *Proceedings of the National Academy of Sciences* 119: e2122084119. https://doi.org/10.1073/pnas.2122084119.

Beck, David. 2000. "Grammatical convergence and the genesis of diversity in the Northwest Coast *Sprachbund*." *Anthropological Linguistics* 42: 147–213.

Beck, David. 2023. "Morphological diffusion and the internal subgrouping of Central Totonac." *Language Dynamics and Change* 14: 1–52.

Beckwith, Brenda R. 1995. *An Evaluation of the Cultural Significance of Plants Used by the Coast Miwok of Marin and Southern Sonoma Counties with Emphasis on Olompoli State Historic Park*. MSc Thesis, California State University, Sacramento, CA.

Bennyhoff, James A. 1994. "Central California Augustine: Implications for Northern California archaeology." In *Toward a New Taxonomic Framework for Central California Archaeology: Essays by James A. Bennyhoff and David A. Fredrickson,* edited by Richard E. Hughes, pp. 65–74. Berkeley, CA: Contributions of the University of California Archaeological Research Facility No. 52.

Berge, Anna. N.d. "Reexamining the Linguistic Prehistory of Aleut." (See section 3.1) Unpublished manuscript in possession of author.

Berge, Anna. 2018. "Re-evaluating the reconstruction of Proto-Eskimo-Aleut." *Journal of Historical Linguistics* 8: 230–272.

Berge, Anna. 2024. "Review of Mid-Holocene Language Connections between Asia and North America. By Michael Fortescue and Edward Vajda (2022)*."* *International Journal of American Linguistics* 90: 130–132.

Bergsland, Knut. *Aleut Dictionary: An Unabridged Lexicon of the Aleutian, Pribilof, and Commander Islands Aleut Language*. Alaska Native Language Center, Fairbanks, AK. 1994.

Bergsland, Knut. 1986. "Comparative Eskimo-Aleut phonology and lexicon." *Journal de la Société Finno-ougrienne* 80: 63–137.

Bettinger, Robert L., and Martin A. Baumhoff. 1982. "The Numic spread: Great Basin cultures in competition." *American Antiquity* 47(3): 485–503.

Blench, Roger 2006. "Archaeology and language: Methods and issues." In *A Companion to Archaeology*, edited by John Bintliff, pp. 52–74. New York: John Wiley and Sons.

Boas, F. *Indian Myths and Legends from the North Pacific Coast of America: A Translation of Franz Boas's 1895 Edition of Indianische Sagen von der Nord Pacifischen Küste Amerikas*. Trans. Dietrich Bertz. Eds. Randy Bouchard and Dorothy Kennedy. Vancouver: Talonbooks. 2002.

Boas, F. 1911. "Introduction." In *Handbook of American Indian languages*, edited by Franz Boas, pp. 5–83. Washington, DC: Bureau of American Ethnology, Bulletin 40.

Boas, Franz, and Pliny Earle Goddard, 1924. "Ts'ets'auts, an Athapascan language from Portland Canal, British Columbia." *International Journal of American Linguistics* 3: 1–35.

Boommelyn, Me'laashne Loren. *Taa-laa-wa Dee-ni' Wee-ya' Tolowa People's Language*. Smith River, CA: The Howonquet Indian Council of the Smith River Rancheria. 2006.

Boraas, Alan and Janet Klein. 1992. "The archaeology of West Point Halibut Cove, Kenai Peninsula, Alaska." *Anthropological Papers of the University of Alaska* 24: 183–204.

Braje, Todd J., Jon M. Erlandson, Torben C. Rick, Loren Davis, Tom Dillehay, Daryl W. Fedje, Duane Froese, Amy Gusick, Quentin Mackie, Duncan McLaren, Bonnie Pitblado, Jennifer Raff, Leslie Reeder-Myers, and Michael R. Waters. 2020. "Fladmark + 40: What have we learned about a potential Pacific Coast peopling of the Americas?" *American Antiquity* 85(1): 1–21.

Breschini, Gary S. *Models of Population Movements in Central California Prehistory*. Salinas, CA: Coyote Press. 1983.

Bright, William. *A Luiseño Dictionary*. Berkeley, CA: University of California Press. 1968.

Brown, Cecil H. *Language and Living Things: Uniformities in Folk Classification and Naming*. New Brunswick, NJ: Rutgers University Press. 1984.

Brown, Cecil H. *Lexical Acculturation in Native American Languages*. New York and Oxford: Oxford University Press. 1999.

Brown, Cecil H. 1981. "Growth and development of folk zoological life-forms in Polynesian languages." *The Journal of the Polynesian Society* 90: 83–110.

Brown, Cecil H. 2011. "The role of Nahuatl in the formation of Mesoamerica as a linguistic area." *Language Dynamics and Change* 1: 171–204.

Brown, Cecil H. 2017. "Evaluating proposals of language genealogical relationship: The Beck-Wichmann-Brown (BWB) system". *Language Dynamics and Change* 7: 252–285.

Brown, Cecil H. 2020. "The Beck-Wichmann-Brown evaluation of lexical comparisons for the Transeurasian proposal." In *The Oxford Guide to the Transeurasian*

Languages, edited by Martine Robbeets and Alexander Savelyev, pp. 735–749. Oxford: Oxford University Press.

Brown, Cecil H., Eric W. Holman, and Søren Wichmann. 2013. "Sound correspondences in the World's languages." *Language* 89: 3–29.

Brown, Cecil H., Eric W. Holman, Søren Wichmann, and Viveka Velupillai. 2008. "Automated classification of the world's languages: A description of the method and preliminary results." *Language Typology and Universals* 61(4): 285–308.

Brown, Cecil H., Søren Wichmann, and David Beck. 2014. "Chitimacha: A Mesoamerican language in the Lower Mississippi Valley." *International Journal of American Linguistics* 80: 425–474.

Brown, H. A. *A Dictionary of Toarioi.* Sydney: University of Sydney. 1968.

Brown, K. J., and R. J. Hebda. 2003. "Coastal rainforest connections disclosed through a late Quaternary vegetation, climate, and fire history investigation from the mountain hemlock zone on southern Vancouver Island, British Columbia, Canada." *Review of Palaeobotany and Palynology* 123: 247–269.

Brown, K. J., A. B. Nielsen, R. J. Fitton, and R. J. Hebda. 2008. "Postglacial evolution and spatial differentiation of seasonal temperate rainforest in Western Canada." *The Holocene* 18(5): 715–727.

Brown, Kendrick J., N. J. R. Hebda, R. J. Hebda, Rich J. Fitton, J. A. Tony Trofymow, and Nicholas Conder. 2022. "Development and wildfire dynamics of dry coastal temperate forests, BC, Canada." *Canadian Journal of Forest Research* 52(10): 1–114. https://doi.org/10.1139/cjfr-2022-0020.

Brown, Thomas J., and Kenneth M. Ames. 2021. "Synthesis and analysis of the Early to Middle Holocene radiocarbon record for the Northern Northwest Coast of North America: A gap in dated components between ~9000 and 8300 cal BP." *The Journal of Island and Coastal Archaeology* 16: 414–437.

Burt, Marina K. *Phonology Wheel.* Oakland, CA: Bloomsbury West, Inc. 1974.

Byrd, Brian F., Adrian R. Whitaker, Patricia J. Mikkelsen, and Jeffrey S. Rosental. 2017. "San Francisco-delta regional context and research design for Native American archaeological resources, CALTRANS District 4." Manuscript on File, California Department of Transportation, District 4, Oakland, CA.

Callaghan, Catherine A. *Bodega Miwok Dictionary.* Berkeley and Los Angeles, CA: University of California Press. 1970.

Callaghan, Catherine A. *Northern Sierra Miwok Dictionary.* Berkeley and Los Angeles, CA: University of California Press. 1987.

Callaghan, Catherine A. *Lake Miwok Dictionary.* Berkeley and Los Angeles, CA: University of California Press. 1965.

Callaghan, Catherine A. *Plains Miwok Dictionary.* Berkeley and Los Angeles, CA: University of California Press. 1984.

Callaghan, Catherine A. *Proto Utian Grammar and Dictionary with Notes on Yokuts.* Berlin: De Gruyter Mouton. 2014.

Campbell, Lyle. *Historical Linguistics: An Introduction.* Cambridge, MA: MIT Press. 1998.

Campbell, Lyle and William J. Poser. *Language Classification: History and Method.* Cambridge: Cambridge University Press. 2008.

Cannon, Aubrey. *The Economic Prehistory of Namu: Patterns in Vertebrate Fauna.* Burnaby: Department of Archaeology and the Institute of Fisheries Analysis: Archaeology Press, Simon Fraser University. 1991.

Carlson, R. L. 1994. "Trade and exchange in prehistoric British Columbia." In *Prehistoric Exchange Systems in North America. Interdisciplinary Contributions to Archaeology,* edited by T. G. Baugh and J. E. Ericson, pp. 307–361. Boston, MA: Springer. https://doi.org/10.1007/978-1-4757-6231-0_11.

Carlson, R. L. 1996. "Introduction to early human occupation in British Columbia." In *Early Human Occupation in British Columbia,* edited by Roy L. Carlson and Luke Dalla Bona, pp. 3–10. Vancouver: UBC Press.

Carlson, R. L. 2014. "Early Namu." In *Early Human Occupation in British Columbia,* edited by R. L. Carlson and L. Dalla Bona, pp. 83–102. Vancouver: UBC Press.

Carlson, R. L., and J. F. Baichtal 2015. "A predictive model for locating Early Holocene archaeological sites based on raised shell-bearing strata in Southeast Alaska, USA." *Geoarchaeology: An International Journal* 30: 120–138.

Clark, Donald W. *Ocean Bay: An Early North Pacific Maritime Culture.* Ottawa: National Museum of Man Mercury Series, Archaeological Survey of Canada Paper No. 86. 1979.

Clark, Donald W. 2001. "Ocean Bay." In *Encyclopedia of Prehistory, Volume 2: Arctic and Subarctic,* edited by Peter N. Peregrine and Melvin Ember, pp. 152–164. New York: Kluwer Academic/Plenum Publishers.

Cohen, Jenny M. 2014. "Paleoethnobotany of Kilgii Gwaay: A 10,700-year-old Ancestral Haida Archaeological Wet Site." MA Thesis, Department of Anthropology, University of Victoria, Victoria.

Collier, Mary E. T., and Sylvia Barker Thalman, eds. *Interviews with Tom Smith and Maria Copa. Isabel Kelly's Ethnographic Notes on the Coast Miwok Indians of Marin and Southern Sonoma Counties, California.* San Rafael, CA: Mapon Occasional Papers Number 6. Miwok Archaeological Preserve of Marin. 1991.

Compton, Brian D. 1993a. "The North Wakashan 'Wild Carrots': Clarification of some ethnobotanical ambiguity in Pacific Northwest Apiaceae." *Economic Botany* 47(3): 297–303.

Compton, Brian D. 1993b. "Upper North Wakashan and Southern Tsimshian ethnobotany: The knowledge and usage of plants and fungi among the Oweekeno, Hanaksiala (Kitlope and Kemano), Haisla (Kitamaat) and Kitasoo peoples of the central and north coast of British Columbia." PhD Dissertation, The University of British Columbia, Vancouver.

Comrie, Bernard. *Language Universals and Linguistic Typology: Syntax and Morphology.* 2nd ed. Oxford: Basil Blackwell. 1989.

Corbett, Debra, and Diane Hanson. *Culture and Archaeology of the Ancestral Unangax̂/Aleut of the Aleutian Islands, Alaska: Unangam Tanangin ilan Unangax̂/Aliguutax̂ Maqax̂singin ama Kadaangim Tanangin Anaĝix̂taqangis.* New York: Springer. 2023.

Cove, John J., and George F. MacDonald, eds. *Tsimshian Narratives I: Tricksters, Shamans and Heroes.* Ottawa: Canadian Museum of Civilization, 1987.

Crawford, Michael H., Rohina C. Rubicz and Mark Zlojutro. 2010. "Origins of Aleuts and the genetic structure of populations of the Archipelago: Molecular and archaeological perspectives." *Human Biology* 82: 695–717.

Croes, Dale R., and Vic J. Kucera. 2017. "Entering the American Continent: The Chehalis River hypothesis." *Journal of Northwest Anthropology* 51: 164–183.

Cruikshank, J. *Do Glaciers Listen? Local Knowledge, Colonial Encounters, and Social Imagination.* Vancouver: UBC Press; Seattle, WA: University of Washington Press. 2005.

Davis, Loren G., and David B. Madsen. 2020. "The coastal migration theory: Formulation and testable hypotheses." *Quaternary Science Reviews* 249. https://doi.org/10.1016/j.quascirev.2020.106605.

DeLancey, Scott and Victor Golla. 1997. "The Penutian hypothesis: Retrospect and prospect." *International Journal of American Linguistics* 63: 171–202.

Deur, Douglas, and Nancy J. Turner, eds. *Keeping It Living: Traditions of Plant Use and Cultivation on the Northwest Coast of North America.* Seattle, WA: University of Washington Press; Vancouver: UBC Press. 2005.

Dinwiddie, J. D. 2014. "The ground slate transition on the northwest coast: Establishing a chronological framework." MSc Thesis, Portland State University, Portland, OR.

Ditidaht Studies Program. *Waacǎkc̓uuʔyk: Ditidaht Dictionary*. ditidahtdictionary.com. 2020.

Dixon, Roland B. and Alfred L. Kroeber. 1919. "Linguistic families of California." *University of California Publications in American Archaeology and Ethnology* 16: 47–118.

Drechsel, Emanuel J. *Mobilian Jargon: Linguistics and Sociohistorical Aspects of a Native American Pidgin.* Oxford: Clarendon Press. 1997.

Dryer, Matthew S. 2013. "Position of Pronominal Possessive Affixes." In *The World Atlas of Language Structures Online,* edited by Matthew S. Dryer and Martin Haspelmath. Leipzig: Max Planck Institute for Evolutionary Anthropology. http://wals.info/chapter/57 (accessed May 27, 2022).

Dryer, Matthew S., and Martin Haspelmath, eds. *The World Atlas of Language Structures Online*. Leipzig: Max Planck Institute for Evolutionary Anthropology. https://wals.info. 2013.

Dumond, Don E. 1984. "Prehistory: Summary." In *Handbook of North American Indians, Volume 5 Arctic*, David Damas, volume editor, pp. 72–93. Washington, DC: Smithsonian Institution.

Dunn, John Asher. *Sm'algyax, A Reference Dictionary and Grammar for the Coast Tsimshian Language.* Seattle, WA: University of Washington Press. 1995.

Eberhard, David M., Gary F. Simons, and Charles D. Fennig, eds. *Ethnologue: Languages of the World.* 25th ed. Dallas, TX: SIL International. http://www.ethnologue.com. 2022.

Edwards, Keri. *Dictionary of Tlingit*. Juneau, AK: Sealaska Heritage Institute. 2009.

Ehret, Christopher. 2012. "Linguistic archaeology." *African Archaeological Review* 29: 109–130.

Emeneau, Murray 1956. "India as a linguistic area." *Language* 32: 3–16.

Enfield, N. J. 2005. "Areal linguistics and Mainland Southeast Asia." *Annual Review of Anthropology* 34: 181–206.

Epps, Patience, and Katherine Bolaños. 2017. "Reconsidering the 'Makú' family of northwest Amazonia." *International Journal of American Linguistics* 83(3): 467–507.

Erlandson, J. M. 2001. "Anatomically modern humans, maritime voyaging, and the Pleistocene colonization of the Americas." In *The First Americans: The Pleistocene Colonization of the New World*, edited by N. G. Jablonski, pp. 1–9. San Francisco, CA: California Academy of Sciences.

Erlandson, Jon M., Michael H. Graham, Bruce J. Bourque, Debra Corbett, James A. Estes, and Robert S. Steneck. 2007. "The kelp highway hypothesis: Marine ecology, the coastal migration theory, and the peopling of the Americas." *Journal of Island and Coastal Archaeology* 2(2): 161–174.

Erlandson, Jon M., Madonna L. Moss, and Mathew Des Lauriers. 2008. "Life on the edge: Early maritime cultures of the Pacific Coast of North America." *Quaternary Science Reviews* 27: 2232–2245.

Eshleman, Jason A., Ripan S. Malhi, John R. Johnson, Frederika A. Kaestle, Joseph Lorenz, and David Glenn Smith. 2004. "Mitochondrial DNA and prehistoric settlements: Native migrations on the western edge of North America." *Human Biology* 76: 55–75.

Fedje, Daryl W., Tina Christensen, H. Josenhans, J. B. McSporran, and J. Strang. 2005. "Millennial tides and shifting shores: Archaeology on a dynamic landscape." In *Haida Gwaii: Human History and Environment from the Time of Loon to the Time of the Iron People*, edited by Daryl W. Fedje and R. W. Mathewes, pp. 164–186. Vancouver: UBC Press.

Fedje, Daryl W., and Q. Mackie. 2005. "Overview of cultural history [of Haida Gwaii]." In *Haida Gwaii: Human History and Environment from the Time of Loon to the Time of the Iron People*, edited by Daryl W. Fedje and R. W. Mathewes, pp. 154–186. Vancouver: UBC Press.

Ferguson, Charles A. 1970. "The Ethiopian language area." *Journal of Ethiopian Studies* 8: 67–80.

Fish, Suzanne K., and Paul R. Fish. 2010 "Monumentality and complex hunter-gatherers in southeast coastal Brazil." In *Monumental Questions: Prehistoric Megaliths, Monuments, and Enclosures*, edited by David Calado, M. Baldia, and M. Boulanger, pp. 231–236. Oxford: BAR International Series 2123.

Fitton, R. J. and R. J. Hebda. 2004. "Late quaternary history of vegetation, climate and fire on south central Vancouver Island, British Columbia, Canada." *Palynology* vol. 28 (1): 243

Fitzhugh, Ben. *The Evolution of Complex Hunter-Gatherers: Archaeological Evidence from the North Pacific*. New York: Kluwer Academic/Plenum Publishers. 2003a.

Fitzhugh, Ben. 2003a. "The evolution of complex hunter-gatherers on the Kodiak Archipelago." In *Hunter-Gatherers of the North Pacific Rim,* edited by J. Habu, J. Savelle, S. Koyama, and H. Hongo, pp. 13–48. SENRI Ethnological Studies, 63. Osaka: National Museum of Ethnology.

Fitzhugh, Ben. 2004. "Colonizing the Kodiak Archipelago: Trends in raw materials use and lithic technologies at the Tanginak Spring site." *Arctic Anthropology* 41(1): 14–40.

Fladmark, Knut. 1979. "Routes: Alternate migration corridors for early man in North America." *American Antiquity* 44: 55–69.

Fortescue, Michael. *Comparative Wakashan Dictionary*. Munich: LINCOM. 2007.

Fortescue, Michael. *Language Relations across Bering Strait*. London and New York: Cassell. 1998.

Fortescue, Michael. 2011. "The relationship of Nivkh to Chukotko-Kamchatkan revisited." *Lingua* 121: 1359–1376.

Fortescue, Michael. 2016. "How the accusative became the relative: A Samoyedic key to the Eskimo-Uralic relationship?" *Journal of Historical Linguistics* 6: 72–92.

Fortescue, Michael, Steven Jacobson, and Lawrence Kaplan. *Comparative Eskimo Dictionary with Aleut Cognates*. 2nd ed. Fairbanks, AK: Alaska Native Language Center. 2010.

Fortescue, Michael, and Edward Vajda. *Mid-Holocene Language Connections between Asia and North America*. Leiden: Brill. 2022.

Frachtenberg, Leo J. *Alsea Texts and Myths*. Washington, DC: Government Printing Office. 1920.

Frachtenberg, Leo J. *Coos Texts*. New York: Columbia University Press. 1913.

Friesen, T. Max, and Owen K. Mason, eds. *The Oxford Handbook of the Prehistoric Arctic*. Oxford: Oxford University Press. https://doi.org/10.1093/oxfordhb/9780199766956.001.0001. 2016.

Georg, Stephan. 2017. "The role of paradigmatic morphology in historical, areal and genealogical linguistics: Thoughts and observations in the margin of paradigm change. In the Transeurasian languages and Beyond (Robbeets and Bisang, eds.)." *Journal of Language Contact* 10(2): 353–355.

Goddard, Ives. *Map of Native Languages and Language Families of North America. (Revised and enlarged edition with additions and corrections.)*. Lincoln, NE: University of Nebraska Press. 1999.

Golla, Victor. *Hupa Language Dictionary: Second Edition*. Hoopa, CA: Hoopa Valley Tribal Council. 1996.

Golla, Victor. 2007. "Linguistic prehistory." In *California Prehistory: Colonization, Culture, and Complexity*, edited by Terry L. Jones, and Kathryn A. Klar, pp. 71–82. New York: Alta Mira Press.

González-Oliver, Angélica, Ernesto Garfias-Morales, David Glenn Smith, and Mirsha Quinto-Sánchez 2017. "Mitochondrial DNA analysis of Mazahua and Otomi Indigenous populations from Estado de México suggests a distant common ancestry." *Human Biology* 89: 195–216.

Gorostiza A, V. Acunha-Alonzo, L. Regalado-Liu, S. Tirado, J. Granados, et al. 2012. "Reconstructing the history of Mesoamerican populations through the study of the mitochondrial DNA control region." *PLoS ONE* 7(9): e44666.

Gould, Richard A. 1978. "Tolowa." In *Handbook of North American Indians*, Volume 8, California, edited by Robert F. Heizer, pp. 128–136. Washington, DC: Smithsonian Institution:

Gray, R., and Q. Atkinson. 2003. "Language-tree divergence times support the Anatolian theory of Indo-European origin." *Nature* 426(6965): 435–439. https://doi.org/10.1038/nature02029.

Gray, R., A. Drummond, and S. Greenhill. 2009. "Language phylogenies reveal expansion pulses and pauses in Pacific settlement." *Science* 323(5913): 479–483. https://doi.org/10.1126/science.1166858.

Greengo, R. E., and R. Houston. 1964 "Excavations at the Marymoor Site, Sammamish River." Unpublished Document on file with SSHPO.

Grier, Colin. 2014. "Landscape construction, ownership and social change in the southern gulf islands of British Columbia." *Canadian Journal of Archaeology/ Journal Canadien d'Archeólogie* 38: 211–249.

Grubb, David McC. *A Practical Writing System and Short Dictionary of Kwakw'ala (Kwakiutl)*. Ottawa: National Museum of Canada. 1977.

Haisla Nation. *Haisla – English Dictionary*. SIL International. www.webonary.org/haisla. 2020.

Hammarström, Harald, Robert Forkel, Martin Haspelmath, and Sebastian Bank. 2022. *Glottolog 4.7*. Leipzig: Max Planck Institute for Evolutionary Anthropology. https://doi.org/10.5281/zenodo.7398962, http://glottolog.org (accessed January 13, 2023).

Harris, Martha Douglas. *History and Folklore of the Cowichan Indians*. Spokane, WA: Marquette Books. 2004. 1901.

Haspelmath, Martin, and Uri Tadmor, eds. *Loanwords in the World's Languages: A Comparative Handbook*. Eds. Martin Haspelmath and Uri Tadmor. Berlin: De Gruyter Mouton. 2009a.

Haspelmath, Martin, and Uri Tadmor. 2009b. "World loanword database (wold)." Leipzig: Max Planck Digital Library. http://wold.clld.org/ (accessed July 18, 2017).

Hausler-Knecht, P. 1993. "Early Prehistory Of The Kodiak Archipelago." Department of Anthropology, Harvard University, Cambridge, MA.

Heath, Jeffrey. *Linguistic Diffusion in Arnhem Land*. Canberra: Australian Institute of Aboriginal Studies. 1978.

Hebda, C., Duncan McLaren, Quentin Mąckie, Daryl W. Fedje, Mikkel Winther Pedersen, Eske Willerslev, Kendrick J. Brown, and Richard J. Hebda. 2022. "Late Pleistocene palaeoenvironments and a possible glacial refugium on northern Vancouver Island, Canada: Evidence for the viability of early human settlement on the northwest coast of North America." *Quaternary Science Reviews* 279: 107388.

Hebda, R. J. 1995. "British Columbia vegetation and climate history with focus on 6 ka BP." *Géographie physique et Quaternaire* 49: 55–79.

Hebda, R. J. 1997. "Late Quaternary paleoecology of Brooks Peninsula." In *Brooks Peninsula: An Ice Age Refugium on Vancouver Island*, edited by R. J. Hebda and J. C. Haggarty. Occasional Paper No. 5, pp. 9.1–9.48. Victoria: BC Parks.

Hebda, R. J., and R. W. Mathewes. 1982. "Postglacial history of cedar and evolution of Northwest Coast native cultures." Paper presented to the 35th annual Northwest Anthropological Conference, Simon Fraser University, Burnaby.

Hebda, Richard J., and S. G. Frederick. 1990. "History of marine resources of the northeast Pacific since the last glaciation." *Transactions of the Royal Society of Canada, Series* 1(1): 319–342.

Hebda, Richard J., Marlow G. Pellatt, Rolf W. Mathewes, Daryl W. Fedje, and Steven Acheson. 2005. "Vegetation history of Anthony Island, Haida Gwaii, and its relationship to climate change and human settlement." In *Haida Gwaii: Human History and Environment from the Time of Loon to the Time of Iron People*, edited by D. W. Fedje and R. W. Mathewes, pp. 59–76. Vancouver: UBC Press.

Heggarty, Paul. 2014. "Prehistory through language and archaeology." In *The Routledge Handbook of Historical Linguistics*, edited by Claire Bowern and Bethwyn Evans, pp. 598–626. New York: Routledge.

Heiltsuk Language & Culture Mobilization Partnership. *Híłzaqvḷa Dictionary*. Sourced from Version 2.1 of the Digital Heiltsuk-English Dictionary © 2007 Heiltsuk Cultural Education Centre, from About - Híłzaqvḷa Dictionary. mothertongues .org. 2021.

Heizer. R. F. *California Indian Linguistic Records: The Mission Indian Vocabularies of Alphonse Pinart*. Berkeley and Los Angeles, CA: University of California Press. 1952.

Heizer, R. F. *California Indian Linguistic Records: The Mission Indian Vocabularies of H.W. Henshaw*. Berkeley and Los Angeles, CA: University of California Press. 1955.

Heizer, R. F. *The Mission Indian Vocabularies of Alphonse Pinart*. Berkeley and Los Angeles, CA: University of California Press. 1951.

Hillis, D., Gustas, R., Pauly, D. W. W. L. Cheung, A. K. Salomon, and I. McKechnie. 2022. "A palaeothermometer of ancient Indigenous fisheries reveals increases in mean temperature of the catch over five millennia." *Environmental Biology of Fishes* 105: 1381–1397.

Hindle, Lonnie, and Bruce Rigsby. 1973. "A short practical dictionary of the Gitksan language." *Northwest Anthropological Research Notes* 7: 1–60.

Hoffman, T., N. Lyons, D. Miller, A, Diaz, A. Homan, S. Huddlestan, and R. Leon, 2016. "Engineered feature used to enhance gardening at a 3800-year-old site on the Pacific Northwest Coast. *Science Advances* 2(12): e1601282. https://doi.org/10.1126/sciadv.1601282.

Hoijer, Harry. 1963. "The Athapaskan languages." In *Studies in the Athapaskan Languages*, edited by Harry Hoijer and others, pp. 1–29. Berkeley and Los Angeles, CA: University of California Press.

Holman, Eric W., Cecil H. Brown, Søren Wichmann, André Müller, Viveka Velupillai, Harald Hammarström, Sebastian Sauppe, Hagen Jung, Dik Bakker, Pamela Brown, Oleg Belyaev, Matthias Urban, Robert Mailhammer, Johann-Mattis List, and Dmitry Egorov. 2011. "Automated dating of the World's language families based on lexical similarity." *Current Anthropology* 52: 841–875.

Holman, Eric W., Søren Wichmann, Cecil H. Brown, Viveka Velupillai, André Müller, and Dik Bakker. 2008. "Explorations in automated language classification." *Folia Linguistica* 42: 331–354.

Hoopa Valley Tribe. *Hupa Language Dictionary*. Hoopa, CA. https://www.slideshare.net/slideshow/hupalanguagedictionary/1818861. 1983.

Hübler, Nataliia, and Simon J. Greenhill. 2023. "Modelling admixture across language levels to evaluate deep history claims." *Journal of Language Evolution* 20: 1–8. https://doi.org/10.1093/jole/lzad002.

Hughes, Cris E., Mary P. Rogers, Amanda C. Owings, Barbara Petzelt, Joycelynn Mitchell, Harold Harry, Theresa Williams, Dena Goldberg, Damian Labuda, David Glenn Smith, Jerome S. Cybulski, and Ripan S. Malhi. 2016. "Genetic structure of First Nation communities in the Pacific Northwest." *Human Biology* 88: 251–263.

Hughes, Richard E. 1992. "California archaeology and linguistic prehistory." *Journal of Anthropological Research* 48: 317–338.

Ives, John W. 2010. "Dene-Yeniseian, migration, and prehistory." In *The Dene-Yeniseian Connection*, edited by James Kari and Ben A. Potter. Special issue, Anthropological Papers of the University of Alaska 5, pp. 324–334. Fairbanks, AK: University of Alaska Department of Anthropology and Alaska Native Language Center.

James, Thomas, E. J. Gowan, Ian Hutchinson, John J. Clague, J. Vaughn Barrie, and Kim W. Conway. 2009. "Sea-level change and paleogeographic reconstructions, southern Vancouver Island, British Columbia, Canada." *Quaternary Science Reviews* 28 (13–14): 1200–1216.

Johnson, John R, Brian M. Kemp, Cara Monroe, and Joseph G. Lorenz. 2012. "A land of diversity: Genetic insights into ancestral origins." In *Contemporary Issues in California Archaeology*, edited by Terry L. Jones and Jennifer E. Perry, pp. 49–72. Walnut Creek, CA: Left Coast Press.

Johnson, John R., and Joseph G. Lorenz. 2006. "Genetics, linguistics, and prehistoric migrations: An analysis of California Indian mitochondrial DNA Lineages." *Journal of California and Great Basin Anthropology* 26: 33–64.

Kaestle, Frederika Ann. "Molecular evidence for prehistoric Native American population movement: The Numic Expansion." PhD Dissertation, University of California, Davis, Davis, CA. 1998.

Kaplan, Lawrence D. 2019. "Bering Strait: A linguistic area bridging two continents." *Journal de la Société des Américanistes* 105: 175–180 .

Kari, James. *Ahtna Athabaskan Dictionary*. Fairbanks, AK: Alaskan Native Languages Center. 1990.

Kari, James. 1978. "Deg Xinag (Ingalik) noun dictionary." Unpublished manuscript.

Kari, James. *Dena'ina Noun Dictionary*. Fairbanks, AK: Alaska Native Language Center. 1977.

Kari, James, and Ben A. Potter, eds. *The Dene-Yeniseian Connection, Special issue, Anthropological Papers of the University of Alaska* 5. Fairbank, AK: University of Alaska. 2010.

Kari, Priscilla Russell. *Plantlore/Dena'ina K'et'una: An Ethnobotany of the Dena'ina People of Southcentral Alaska*. Denver, CO: University Press of Colorado and Fairbanks, AK: University of Alaska Press. 2020.

Kaufman, Darrell S., Yarrow L. Axford, Andrew C.G. Henderson, Nicholas P. McKay, W. Wyatt Oswald, Casey Saenger, R. Scott Anderson, Hannah L. Bailey, Benjamin Clegg, Konrad Gajewski, Feng Sheng Hu, Miriam C. Jones, Charly Massa, Cody C. Routson, Al Werner, Matthew J. Wooller, and Zicheng Yu. 2016.

"Holocene climate changes in eastern Beringia (NW North America) – A systematic review of multi-proxy evidence." *Quaternary Science Reviews* 147: 312–339.

Kaufman, Terrence. *Proyecto de alfabetos y ortografías para escribir las lenguas mayances*. Guatemala: PROYECTO LINGUISTICO FRANCISCO MARROQUIN. 1970.

Kessler, Brett. *The Significance of Word Lists: Statistical Tests for Investigating Historical Connections between Languages*. Stanford, CA: CSLI Publications. 2001.

Kinkade, M. Dale. *Upper Chehalis Dictionary*. Missoula, MT: The University of Montana. 1991.

Kinkade, M. Dale. 1989. "Prehistory of Salishan languages." Paper presented at the 8th Annual Meeting of the American Anthropological Association, Washington, DC, November 17, 1989.

Kjær, Kurt H., Anders A. Bjørk, Kristian K. Kjeldsen, Eric S. Hansen, Camilla S. Andresen, Marie-Louise Siggaard-Andersen, Shfaqat A. Khan, Anne Sofie Søndergaard, William Colgan, Anders Schomacker, Sarah Woodroffe, Svend Funder, Alexandra Rouillard, Jens Fog Jensen, and Nicolaj K. Larsen. 2022. "Glacier response to the Little Ice Age during the Neoglacial cooling in Greenland." *Earth-Science Reviews* 227. https://doi.org/10.1016/j.earscirev.2022.103984.

Klinkenberg, Brian, ed. *E-Flora BC: Electronic Atlas of the Flora of British Columbia [eflora.bc.ca]. Lab for Advanced Spatial Analysis*. Vancouver: Department of Geography, The University of British Columbia. 2023.

Kopperl, Robert E. 2012. "Chronology of the Ocean Bay Tradition on Kodiak Island, Alaska: Stratigraphic and radiocarbon analysis of the Rice Ridge Site (KOD-363)." *Alaska Journal of Anthropology* 10(1&2): 17–35.

Kopperl, Robert E. 2003. "Cultural complexity and resource intensification on Kodiak Island, Alaska," PhD Dissertation, Department of Anthropology, University of Washington, Seattle, WA.

Krauss, Michael E. *Eyak Dictionary*. Fairbanks, AK: University of Alaska and Cambridge, MA: Massachusetts Institute of Technology. 1963–1970.

Krauss Michael E. 1964. "Proto-Athapaskan-Eyak and the problem of Na-Dene: The phonology." *International Journal of American Linguistics* 30: 118–131.

Krauss, Michael E. 1974. "Minto Nenana Athapaskan noun dictionary." Unpublished manuscript (Resource Details). Alaska Native Language Archive. uaf.edu.

Krauss, Michael E., and Victor Golla. 1981. "Northern Athapaskan languages." In *Handbook of North American Indians: Subarctic*, edited by June Helm, pp. 67–85. Washington, DC: Smithsonian Institution.

Kroeber, A. L. *The Chumash and Costanoan Languages*. Berkeley, CA: The University Press. 1910.

Kuhnlein, H. V., and N. J. Turner. *Traditional Plant Foods of Canadian Indigenous Peoples: Nutrition, Botany and Use*. London: Routledge Revivals, Taylor & Francis, CRC Press. [Originally published 1991]. 2020.

Kuipers, Aert H. *Salish Etymological Dictionary*. Missoula, MT: University of Montana. 2002.

Kuipers, Aert H. *The Squamish Language*. The Hague: Mouton & Co. 1967.

Lacourse, Terri, J. Michelle Delepine, Elizabeth H. Hoffman, and Rolf W. Mathewes. 2012. "A 14,000 year vegetation history of a hypermaritime island on the outer Pacific coast of Canada based on fossil pollen, spores, and conifer stomata." *Quaternary Research* 78: 572–582.

Lamb, Sydney M. 1958. "Linguistic prehistory in the Great Basin." *International Journal of American Linguistics* 24: 95–100. Franz Boas Centennial Volume (April 1958).

Lang, George. *Making Wana: The Genesis of Chinook Jargon.* Vancouver and Toronto: UBC Press. 2008.

Lawrence, Irma. *Haida Dictionary.* Fairbanks, AK: The Alaska Native Language Center. 1977.

Lee, E. S. 1966. "A theory of migration." *Demography* 3: 47–57.

Leer, Jeff. *Conversational Alutiiq Dictionary.* Fairbanks, AK: Alaska Native Language Center. 1978a.

Leer, Jeff. *Kodiak Alutiiq.* Fairbanks, AK: Alaska Native Language Center. 1978b.

Lefferts, H. L., Jr. 1977. "Frontier demography: An introduction." In *The Frontier, Comparative Studies,* edited by D. H. Miller and J. O. Steffen, pp. 33–35. Norman, OK: University of Oklahoma Press.

Lepofsky, D., and C. G. Armstrong. 2018. "Foraging new ground: Documenting ancient resource and environmental management in Canadian archaeology." *Canadian Journal of Archaeology* 42: 57–73.

Lepofsky, D., K. Lertzman, D. Hallett, and R. Mathewes. 2005. "Climate change and culture change on the Southern Coast of British Columbia 2400-1200 B.P.: An hypothesis." *American Antiquity* 70: 267–293.

Lepofsky, D., B. Letham, M. Ritchie, and C. G. Armstrong. 2024. "Placemaking on the Northwest Coast." In *The Oxford Handbook of Island and Coastal Archaeology*, edited by S. Fitzpatrick and J. Erlandson. Oxford Academic. oup.com.

Lepofsky, Dana, David Schaepe, Anthony Graesch, Michael Lenert, Patricia Ormerod, Keith Carlson, Jeanne Arnold, Michael Blake, Patrick Moore, and John Clague. 2009. "Exploring Stó:lō-Coast Salish interaction and identity in ancient houses and settlements in the Fraser Valley." *British Columbia American Antiquity* 74: 595–626.

Letham, B., I. Dundas, J. Earnshaw, B. Eshom, S. Fisher, G. Fisher, S. Greening, E. Letham, J. Reece, R. Robinson, and I. Sellers. 2022. "The Moore Islands Project: Deep-Time Indigenous history and landscape change on the outer Northwest Coast of North America." *Antiquity* 96(386): 455–462. https://doi.org/10.15184/aqy.2022.10.

Letham, B., D. Lepofsky, and S. Greening. 2021. "A post-glacial relative sea level curve for the Central Douglas Channel area, British Columbia, Canada." *Quaternary Science Reviews* 263: 106991.

Levenshtein, Vladimir I. 1966. "Binary codes capable of correcting deletions, insertions, and reversals." *Soviet Physics Doklady* 10: 707–710.

Levy, Richard L. 1978. "The linguistic prehistory of California: A processual view." Paper presented at the annual meeting of the Society of American Archaeology, Tucson, AZ.

Lewis, David. *Athapaskan Indians*. Oregon Encyclopedia, Oregon Historical Society. Athapaskan Indians. oregonencyclopedia.org. 2023.

Lightfoot, Kent G. 1997. "Cultural construction of coastal landscapes: A Middle Holocene perspective from San Francisco Bay." In *Archaeology of the California Coast during the Middle Holocene*, edited by Jon Erlandson and Michael Glassow, pp. 129–141. Los Angeles, CA: Institute of Archaeology, UCLA.

Lightfoot, Kent G., Rob Q. Cuthrell, Mark G. Hylkema, Valentin Lopez, Diane Gifford-Gonzalez, Roberta A. Jewett, Michael A. Grone, Gabriel M. Sanchez, Peter A. Nelson, Alec J. Apodaca, Ariadna Gonzalez, Kathryn Field, Jordan F. Brown, Alexii Sigona, and Paul V. A. Fine. 2021. "The eco-archaeological investigation of Indigenous stewardship practices on the Santa Cruz coast." *Journal of California and Great Basin Anthropology* 41(2): 185–204.

Lightfoot, Kent G., and Valentin Lopez. 2013. "The study of Indigenous management practices in California: An Introduction." *California Archaeology* 5(2): 209–219.

Lightfoot, Kent G., and Edward M. Luby. 2002. "Late Holocene in the Greater San Francisco Bay Area: Temporal trends in the use and abandonment of shell mounds in the East Bay." In *Catalysts to Complexity: The Late Holocene on the California Coast*, edited by Jon Erlandson, and Terry Jones, pp. 263–281. Los Angeles, CA: Costen Institute of Archaeology, UCLA.

Lightfoot, Kent G., Edward M. Luby, Rob Q. Cuthrell, Michael Grone, Gabriel M. Sanchez, Alec J. Apodaca, and Ariadna Gonzalez. *New Insights from Legacy Collections: Revisiting CA-ALA-207, A Large California Shell Mound.* Washington, DC: Final Report, National Science Foundation, 2022.

Lightfoot, Kent G., Antoinette Martinez, and Ann Schiff. 1998. "Daily practice and material culture in pluralistic social settings: An archaeological study of culture change and persistence from Fort Ross, California." *American Antiquity* 63(2): 199–222.

List, Johann-Mattis. 2019. "Automatic inference of sound correspondence patterns across multiple languages." *Computational Linguistics* 45: 137–161.

List, Johann-Mattis, Simon J. Greenhill, and Russell D. Gray. 2017. "The potential of automatic word comparison for historical linguistics." *PLoS ONE* 12: e0170046. https://doi.org/10.1371/journal.pone.0170046.

Lorenz, J., and D. G. Smith. 1996. "Distribution of four founding mtDNA haplogroups among native North Americans." *American Journal of Physical Anthropology* 101: 307–323.

Losey, R. 2003. "Sea-level history and the archaeology of the Northern Oregon Coast: A case study from Netarts Bay." *Journal of Northwest Anthropology* 37(2): 239–259.

Ludwin, R. S., R. Dennis, D. Carver, A. D. McMillan, R. Losey, J. Clague, C. Jonientz-Trisler, J. Bowechop, J. Wray, and K. James. 2005. "Dating the 1700 Cascadia earthquake: Great coastal earthquakes in native stories." *Seismological Research Letters* 76(2): 140–147.

MacDonald, B. L. 2008. "Ochre procurement and distribution on the central coast of British Columbia." MA Thesis, McMaster University, Hamilton.

Mackie, A., D. W. Fedje, Q. Mackie, and M. Florian. 2003. "Early Holocene wood artifacts from the Kilgii Gwaay site, Southern Haida Gwaii." Paper presented at the 10th International Wetland Archaeology Research Project Conference, South Puget Sound Community College, Olympia, Washington, DC, 1–5 April.

Mackie, Q., Daryl Fedje, Duncan McLaren, Nicole Smith, and Iain McKechnie. 2011. "Early environments and archaeology of Coastal British Columbia." In *Trekking the Shore: Changing Coastlines and the Antiquity of Coastal Settlement*, edited by N. F. Bicho, J. A. Haws, and L. G. Davis, pp. 51–103. New York: Springer.

Magne, Martin P. R., and R. G. Matson. 2010. "Moving on: Expanding perspectives on Athapaskan migration." *Canadian Journal of Archaeology* 34: 212–239.

Malayalam Dictionary. Lexilogos. Google Traduction. https://translate.google.com// (accessed December 20, 2021).

Malhi, R. S., H. M. Mortensen, J. A. Eshleman, Jason A. Kemp, Brian M. Lorenz, Joseph G. Kaestle, Frederika A. Johnson, John R. Gorodezky, Clara Smith, and David Glenn. 2004. "Native American mtDNA prehistory in the American Southwest". *American Journal of Physical Anthropology* 120: 108–124.

Malhi, Ripan Singh, Angelica Gonzalez-Oliver, Kari Britt Schroeder, Brian M Kemp, Jonathan A. Greenberg, Solomon Z. Dobrowski, David Glenn Smith, Andres Resendez, Tatiana Karafet, Michael Hammer, Stephen Zegura, and Tatiana Brovko. 2008. "Distribution of Y chromosomes among Native North Americans: A study of Athapaskan population history." *American Journal of Physical Anthropology* 137(4): 412–424.

Mallory, Fintan. 2021. "The case against linguistic palaeontology." *Topoi* 40: 273–284. https://doi.org/10.1007/s11245-11020-09691-11245.

Martindale, Andrew, and Susan Marsden. 2003. "Defining the Middle Period: (3500 BP to 1500 BP) in Tsimshian history through a comparison of archaeological and oral records." *BC Studies* 138/139: 13–50.

Maschner, Herbert D. G. 2016. "Archaeology of the Eastern Aleut Region." In *The Oxford Handbook of the Prehistoric Arctic,* edited by T. Max Friesen and Owen K. Mason, pp. 323–348. Oxford: Oxford University Press.

Mason, J. Alden. *The Language of the Salinan Indians*. Berkeley, CA: University of California Press. 1923.

Mason, J. Alden. *The Mutsun Dialect of Costanoan Based on the Vocabulary of De La Cuesta*. Berkeley, CA: University of California Press. 1916.

Mathewes, R. W. 1973. "A palynological study of postglacial vegetation changes in the University Research Forest, southwestern British Columbia." *Canadian Journal of Botany* 51: 2085–2103.

Matson, R. G., and G. Coupland. *The Prehistory of the Northwest Coast*. Toronto: Academic Press. 1995.

McKechnie, Iain, and Rebecca J. Wiggen. 2011. "Toward a historical ecology of pinniped and sea otter hunting traditions on the coast of Southern British Columbia." In *Human Impacts on Seals, Sea Lions, and Sea Otters: Integrating Archaeology and Ecology in the Northeast Pacific*, edited by Todd J. Braje and Torben C. Rick, pp. 129–166. Oakland, CA: University of California Press. https://doi.org/10.1525/california/9780520267268.003.0007.

McLaren, D. 2008. "Sea level change and archaeological site locations on the Dundas Island Archipelago of North Coastal British Columbia." PhD Dissertation, University of Victoria, Victoria.

McLaren, D. 2017. "Long-term human occupation of the Stave Watershed: Subsurface testing, radiocarbon Dates and archaeological material." In *Archaeology of the Lower Fraser River Region*, edited by Michael Rousseau and Roy L. Carlson, pp. 149–158. Burnaby: Simon Fraser University, Archaeology Press.

McLaren, D., D. W. Fedje, A. Dyck, Q. Mackie, A. Gauvreau, and J. Cohen. 2018. "Terminal Pleistocene epoch human footprints from the Pacific coast of Canada." *PLoS ONE* 13(3): e0193522. https://doi.org/10.1371/journal.pone.0193522.

McLaren, D., D. W. Fedje, Q. Mackie, L. G. Davis, J. Erlandson, A. Gauvreau, and C. Vogelaar. 2020. "Late Pleistocene archaeological discovery models on the Pacific coast of North America." *PaleoAmerica* 6: 43–63.

McLaren, D., Kathleen L. Hawes, Tyler Graham, Alisha Gauvreau, Jenny M. Cohen, and Dale R. Croes. 2019. "Early and Middle Holocene waterlogged materials from archaeological sites on the Central Coast of British Columbia." In *Waterlogged. Examples and Procedures for Northwest Coast Archaeologists*, edited by K. Bernick, pp 169–192. Pullman, WA: Washington State University Press.

McLaren, D., F. Rahemtulla, Gitla (E. White), and D. W. Fedje. 2015. "Prerogatives, sea level, and the strength of persistent places: Archaeological evidence for long-term occupation of the Central Coast of British Columbia." *BC Studies* 187: 155–306.

McMillan, Alan D. 2003. "Reviewing the Wakashan migration hypothesis." In *Emerging from the Mist: Studies in Northwest Coast Culture History*, edited by R. G. Matson, Gary Coupland, and Quentin Mackie, pp. 244–259. Vancouver: University of British Columbia Press.

McMillan, Alan D. 1998. "Changing views of Nuu-Chah-Nulth culture history: Evidence of population replacement in Barkley Sound." *Canadian Journal of Archaeology* 22: 5–18.

McNiven, Ian J. 2013. "Ritualized middening practices." *Journal of Archaeological Method and Theory* 20: 552–587.

Merriwether, D. A., F. Rothhammer, and R. E. Ferrell. 1995. "Distribution of the four founding haplotypes in Native Americans suggest a single-wave migration for the New World." *American Journal of Physical Anthropology* 98: 411–430.

Meyer, Jack, and Philip Kaijankoski. 2017 "Discovering sites: Geoarchaeological approaches to site sensitivity and predictive modeling." In *San Francisco-Delta Regional Context and Research Design for Native American Archaeological Resources, CALTRANS District 4*, edited by Brian F. Byrd, Adrian R. Whitaker, Patricia J. Mikkelsen, and Jeffrey S. Rosental, pp. 4.1–4.13. Oakland, CA: Manuscript on File, California Department of Transportation, District 4.

Meyer, Jack, and Jeffrey S. Rosenthal. 1997 "Archaeological and geoarchaeological investigations at eight prehistoric sites in the Los Vaqueros Reservoir Area, Contra Costa County, California." Archaeological Studies Center, Sonoma State University Academic Foundation, Inc., Rohnert Park, CA.

Milliken, Randall, Richard T. Fitzgerald, Mark G. Hylkema, Randy Gorza, Tom Origer, David G. Bieling, Alan Leventhal, Randy S.Wiberg, Andrew Gottsfield, Donna Gillette, Viviana Bellifemine, Eric Strother, Robert Cartier, and David A. Fredrickson. 2007. "Punctuated culture change in the San Francisco Bay Area." In *California Prehistory*, edited by Terry L. Jones, and Kathryn A. Klar, pp. 99–123. Walnut Creek, CA: Alta Mira Press.

Mitchell, Donald H. 1990. "Prehistory of the coasts of Southern British Columbia and Northern Washington." In *Handbook of North American Indians. Vol 7. The Northwest Coast*, edited by Wayne Suttpes, pp. 340–358. Washington, DC: Smithsonian Institution Press.

Mitchell, Donald H. 1988. "Changing patterns of resource use in the prehistory of Queen Charlotte Strait, British Columbia." In *Prehistoric Economies of the Pacific Northwest Coast*, edited by B. L. Isaac, pp. 245–290. Greenwich: JAI Press.

Monroe, Cara Rachelle. "Correlating biological relationships, social inequality, and population movement among prehistoric California foragers: Ancient human DNA analysis from CA-SCL-38 (Yukisma site)." PhD Dissertation, University of California, Santa Barbara, Santa Barbara, CA. 2014.

Moratto, Michael J. *California Archaeology*. Orlando, FL: Academic Press. 1984.

Morgan, Christopher. 2010. "Numic expansion in the Southern Sierra Nevada." *Journal of California and Great Basin Anthropology* 30: 157–174.

Moss, Madonna L. *Northwest Coast: Archaeology as deep history*. Washington, DC: Society for American Archaeology Press. 2011.

Moss, Madonna L., Amy Shannon, Brittany Falconer, Scott Blumenthal, Jensen Wainwright, Elizabeth McGuire, and Molly R. Casperson. 2021. "Alutiiq ancestors use of birds during the Ocean Bay Period at Rice Ridge (49-KOD-363), Kodiak Island, Alaska." *Arctic Anthropology* 58(1): 1–33.

Moss, Madonna L., Dorothy M. Peteet, and Cathy Whitlock. 2007. "Mid-Holocene culture and climate on the Northwest Coast of North America." In *Climate Change and Cultural Dynamics: A Global Perspective on Mid-Holocene Transitions*, edited by David G. Anderson, Kirk A. Maasch, and Daniel H. Sandweiss, pp. 491–529. San Diego, CA: Elsevier and Academic Press.

Munroe, Pamela, and Dieynaba Gaye. *Ay Baati Wolof: A Wolof Dictionary*. Los Angeles, CA: University of California. 1997.

Müller, André, Viveka Velupillai, Søren Wichmann, Cecil H. Brown, Eric W. Holman, Sebastian Sauppe, Pamela Brown, Harald Hammarström, Oleg Belyaev, Johann-Mattis List, Dik Bakker, Dmitri Egorov, Matthias Urban, Robert Mailhammer, Matthew S. Dryer, Evgenia Korovina, David Beck, Helen Geyer, Pattie Epps, Anthony Grant, and Pilar Valenzuela. 2021. *ASJP World Language Trees of Lexical Similarity: Version 5 (October 2021)*. https://asjp.clld.org/download.

Nater, H. F. *A Concise Nuxalk English Dictionary*. Gatineau: Canadian Museum of Civilization. 1990.

Nichols, Johanna. *Linguistic Diversity in Space and Time*. Chicago, IL: The University of Chicago Press. 1992.

Nichols, Johanna. 2010. "Proof of Dene-Yeniseic relatedness." In, *The Dene-Yeniseian Connection,* edited by James Kari and Ben A. Potter, pp. 266–278. Fairbanks, AK: Alaska Native Language Center.

Nichols, Johanna, and Balthasar Bickel. 2013a. "Locus of marking in the clause." In *The World Atlas of Language Structures Online,* edited by Matthew S. Dryer and Martin Haspelmath. Leipzig: Max Planck Institute for Evolutionary Anthropology. http://wals.info/chapter/23 (accessed May 25, 2022).

Nichols, Johanna, and Balthasar Bickel. 2013b. "Locus of Marking in Possessive Noun Phrases." In *The World Atlas of Language Structures Online,* edited by Matthew S. Dryer and Martin Haspelmath. Leipzig: Max Planck Institute for Evolutionary Anthropology. http://wals.info/chapter/24 (accessed May 25, 2022).

Ó Siochfhradha. M. *Learner's English-Irish Dictionary*. Dublin: COMHLACHT OIDEACHAIS NA HÉIREANN, TTA. n.d.

The Ocean Rowing International. 2010. "Completed ocean rows in chronological order." oceanrowing.com. (accessed January 8, 2022).

O'Rourke, D. H., M. G. Hayes, and S. W. Carlyle. 2000. "Spatial and temporal stability of mtDNA haplogroup frequencies in Native North America." *Human Biology* 72: 15–34.

Oswalt, Robert. *Kashaya Dictionary*. Webonary.org. 2018.

Palmer, Hannah M., Veronica Padilla Vriesman, Caitlin M. Livsey, Carina R. Fish, and Tessa M. Hill. 2023. "Holocene climate and oceanography of the coastal Western United States and California current system." *Climate of the Past* 19: 199–232. https://doi.org/10.5194/cp-19-199-2023.

Peñaloza-Espinosa, R. I., D. Arenas-Aranda, R. M. Cerda Flores, et al. 2007. "Characterization of mtDNA haplogroups in 14 Mexican Indigenous populations." *Human Biology* 79: 313–320.

Peteet, Dorothy M., Jonathan E. Nichols, and Daniel H. Mann. 2019. "Holocene vegetation, climate, and carbon history on western Kodiak Island, Alaska." *Frontiers of Earth Science. Section on Quaternary Science, Geomorphology and Paleoenvironment* 7. https://doi.org/10.3389/feart.2019.00061.

Piispanen, Peter Sauli. 2023. "The influence of Chukchi, an ancient Lingua Franca, onto the Yukaghir languages and dialects." Unpublished manuscript in possession of the paper's author.

Potter, Ben A. 2010. "Archaeological patterning in Northeast Asia and Northwest North America: An examination of the Dene-Yeniseian hypothesis." In *The Dene-Yeniseian Connection*, edited by James Kari and Ben A. Potter. Special issue, Anthropological Papers of the University of Alaska 5, pp. 138–167. Fairbanks, AK: University of Alaska.

Powell, J. V. and Fred Woodruff, Sr. 1976. "Quileute Dictionary." Vol. 10, No. 1, Part 2. Moscow, ID: Northwest Anthropological Research Notes.

Pullum, Geoffrey K., and William A. Ladusaw. *Phonetic Symbol Guide*. Chicago, IL: The University of Chicago Press. 1986.

Raff, Jennifer A., Deborah A. Bolnick, Justin Tackney, and Dennis H. O'Rourke. 2011. "Ancient DNA perspectives on American colonization and population history." *American Journal of Physical Anthropology* 146: 503–514.

Rama, T., and S. Wichmann. 2020. "A test of Generalized Bayesian Dating: A new linguistic dating method." *PLoS ONE* 15(8): e0236522. https://doi.org/10.1371/journal.pone.0236522.

Rask, Rasmus. 1820 (see Thalbitzer 1921).

Redmond, Jennifer. 2023. "Placing CA-SFR-220 in context: An introduction to the identification of a 7,500-year-old buried site." Paper presented at the 2023, March 16–19 Annual Meeting for the Society for California Archaeology, Oakland, CA.

Rehg, Kenneth L., and Damian G. Sohl. *Ponapean-English Dictionary*. Honolulu, HI: The University Press of Hawaii. 1979.

Ringe, Donald A. 2002. "Indo-European and its closest relatives: The Eurasiatic language family, Vol 1: Grammar by Joseph H. Greenberg" (Review). *Journal of Linguistics* 38: 415–429.

Robbeets, Martine, and Alexander Savelyev, eds. *The Oxford Guide to the Transeurasian Languages*. Oxford: Oxford University Press. 2020.

Robins, Robert. *The Yurok Language, Grammar, Text, Lexicon*. Berkeley, CA: University of California Press. 1958.

Rosenthal, Jeffrey S. *Archaeological Investigations at CA-CCO-18/548 along Marsh Creek, Brentwood, California, 1946-2008*, Manuscript On File, Far Western Anthropological Research Group, Inc., Davis, CA 2010.

Rosenthal, Jeffrey S., and Jack Meyer. 2004a. "Cultural resources inventory of Caltrans District 10 Rural Conventional Highways," Vol. III: Geoarchaeological Study, Landscape Evolution and the Archaeological Record of Central California. Far Western Anthropological Group Inc., Davis, CA. Prepared for California Department of Transportation, District 10, Stockton.

Rosenthal, Jeffrey S., and Jack Meyer. *Landscape Evolution and the Archaeological Record: A Geoarchaeological Study of the Southern Santa Clara Valley and Surrounding Region.* Davis, CA: Center for Archaeological Research at Davis Publication 14. University of California, Davis. 2004b.

Rouse, Irving. *Migrations in Prehistory: Inferring Population Movement from Cultural Remains.* New Haven, CT: Yale University Press. 1986.

Rouse, Irving. 1958. "The inference of migrations from anthropological evidence." In *Migrations in New World Culture History*, Vol Social Science Bulletin No. 27, University of Arizona Bulletin, Volume 29, Number 2, edited by Raymond H. Thompson, pp. 63–68. Tucson, AZ: University of Arizona Press.

Routson, Kanin J., G. M. Volk, C. M. Richards, S. E. Smith, G. P. Nabhan, and V. Wyllie de Echeverria. 2012. "Genetic variation and distribution of Pacific crabapple." *Journal of the American Society of Horticultural Science* 137(5): 325–332.

Rubicz, Rohina, Theodore G. Schurr, Paul L. Babb, and Michael H. Crawford. 2003. "Mitochondrial DNA variation and the origins of the Aleuts." *Human Biology* 75: 809–835.

Saitou, N., and M. Nei. 1987. "The neighbor-joining method: A new method for reconstructing phylogenealogical trees." *Molecular Biology and Evolution* 4: 406–425. https://doi.org/10.1093/oxfordjournals.molbev.a040454.

Saltonstall, Patrick G., Amy F. Steffian, and Amanda A. Lancaster. *The Alutiiq Technological Inventory: An Account of the Manufacturing Industries, Tools, and Raw Materials Found in Ancestral Alutiiq Archaeological Sites of the Kodiak Archipelago*. Kodiak, AK: Alutiiq Museum and Archaeological Repository. 2021.

Sapir, Edward. 1929. "Central and North American languages." *Encyclopaedia Brittanica* 5: 138–141.

Sassaman, Kenneth E. 2004. "Complex hunter-gatherers in evolution and History: A North American perspective." *Journal of Archaeological Research* 12(3): 227–280.

Sawyer, Jesse O. *English-Wappo Vocabulary*. Berkeley, CA: University of California Press. 1965.

Sawyer, Jesse O., and Alice Schlichter. *Yuki Vocabulary*. Berkeley, CA: University of California Press. 1984.

Schalk, R. *The Evolution and Diversification of Native Land Use Systems on the Olympic Peninsula*. Washington, DC: National Park Service. 1988.

Schurr, T. G., R. Sukernik, Y. Starikovskaya, et al. 1999. "Mitochondrial DNA variation in Koryaks and Itel'men: Population replacement in the Okhotsk Sea-Bering Sea region during the Neolithic." *American Journal of Physical Anthropology* 108: 1–39.

Scott, G. Richard, and Dennis O'Rourke. 2010. "Genes across Beringia: A physical anthropological perspective on the Dene-Yeniseian hypothesis." In *The Dene-Yeniseian Connection*, edited by James Kari and Ben A. Potter, pp. 119–137. Special issue, Anthropological Papers of the University of Alaska 5. Fairbanks, AK: University of Alaska.

Shaul, David L. 1995. "The Huelel (Esselen) language." *International Journal of American Linguistics* 61: 191–239.

Shipley, William. 1983–1984. "Penutian among the ruins: A personal assessment." *Journal of the Steward Anthropological Society* 15: 59–63.

Shugar, D. H., I. J. Walker, O. B. Lian, J. B. R. Eamer, C. Neudorf, D. McLaren, and D. W. Fedje. 2014. Post-glacial sea-level change along the Pacific coast of North America. *Quaternary Science Reviews* 97: 170–192.

Smith, Silvia E., M. Geoffrey Hayes, Graciela S. Cabana, Chad Huff, Joan Brenner Coltrain, and Dennis H. O'Rourke. 2009. "Inferring population continuity versus replacement with aDNA: A cautionary tale from the Aleutian Islands." *Human Biology* 81: 407–426.

Snoek, Conor. 2015. "The lexical semantics of Athapaskan anatomical terms: A historical-comparative study." Unpublished PhD Thesis, University of Alberta, Alberta.

Spence, Michael W. 2005. "A Zapotec diaspora network in Classic-Period Central Mexico." In *The Archaeology of Colonial Encounters: Comparative Perspectives*, edited by Gil J. Stein, pp. 173–205. Santa Fe, NM: School of American Research Press.

Steffian, Amy F., E. P. Eufemio, and Patrick G. Saltonstall. 2002. "Early sites and microblade technologies from the Kodiak Archipelago." *Anthropological Papers of the University of Alaska* 2(1): 1–38.

Steffian, Amy F., and Patrick G. Saltonstall. In press (expected 2024). "Cuumillat Ancestors." In *Imaken Ima'ut—From the Future to the Present: Seventy-Five*

Hundred Years of Kodiak, edited by Amy F. Steffian, pp. 31–73. Kodiak, AK: Alutiiq Museum.

Steffian, Amy F., Patrick, G. Saltonstall, and R. Kopperl. 2006. "Expanding the Kachemak: Surplus production and the development of multi-season storage in Alaska's Kodiak Archipelago." *Arctic Anthropology* 43: 93–129.

Stein, Gil J. 2005. "Political economy of Mesopotamian colonial encounters." In *The Archaeology of Colonial Encounters: Comparative Perspectives*, edited by Gil J. Stein, pp. 143–171. Santa Fe, NMe: School of American Research Press.

Stevens, Michelle L. 2004. Ethnoecology of Selected California Wetland Plants. *Fremontia* 32(4): 7–15.

Stewart, Mark Fleisher. 1976. "Clallam: A study in Coast Salish ethnolinguistics." Unpublished PhD Dissertation, Washington State University, Pullman, WA.

Stoneking, Mark. 2016. *An Introduction to Molecular Anthropology.* Hoboken, NJ: Wiley-Blackwell. 2016.

Stonham, John. *A Concise Dictionary of the Nuuchahnulth Language of Vancouver Island.* Lewiston, NY: The Edwin Mellen Press. 2005.

Sutton, Mark Q. 1991. "Approaches to linguistic prehistory." *North American Archaeologist* 12(4): 303–324.

Sutton, Mark Q. 2014. "Warfare and expansion: An Ethnohistoric perspective on the Numic spread." In *Violence and Warfare among Hunter-Gatherers*, edited by Mark W. Allen and Terry L. Jones, pp. 149–167. New York: Taylor and Francis Group.

Swadesh, Morris. 1952. "Lexicostatistic dating of prehistoric ethnic contacts." *Proceedings of the American Philosophical Society* 96: 452–463.

Swadesh, Morris. 1953a. "Mosan I: A problem of remote common origin." *International Journal of American Linguistics* 19: 26–44.

Swadesh, Morris. 1953b. "Mosan II: Comparative vocabulary." *International Journal of American Linguistics* 19: 223–236.

Swadesh, Morris. 1953c. "Salish-Wakashan lexical comparisons noted by Boas." *International Journal of American Linguistics* 19: 290–299.

Swadesh, Morris. 1955b. "Chemakum lexicon compared with Quileute." *International Journal of American Linguistics* 21: 60–72.

Swadesh, Morris. 1955a. "Towards greater accuracy in lexicostatistic dating." *International Journal of American Linguistics* 21: 121–137.

Takahashi, M. *Romanized English-Japanese Japanese-English Dictionary.* Tokyo: Taiseido Shobo Co. 1953.

Teeter, Karl V., and John D. Nichols. *Wiyot Handbook II, Interlinear Translation and English Index.* Winnipeg: Algonquian and Iroquoian Linguistics, Memoir 11. 1993.

Tenenbaum, Joan M. *Nondalton Tanaina Noun Dictionary.* Fairbanks, AK: Alaska Native Language Center. 1975.

Thalbitzer, William. 1921. "The Aleutian language compared with Greenlandic: A manuscript by Rasmus Rask, Dating from 1820, Now in the Royal Library at Copenhagen." *International Journal of American Linguistics* 2: 1/2.40–57.

Thalman, Sylvia Barker. *The Coast Miwok Indians of the Point Reyes Area.* Point Reyes Station, CA: Point Reyes National Seashore Association. 1993.

Thomason, Sarah Grey. 2000. "Linguistic areas and language history." *Studies in Slavic and General Linguistics* 28: 311–327.

Thomason, Sarah Grey, and Terrence Kaufman. *Language Contact, Creolization, and Genealogical Linguistics*. Berkeley, CA: University of California Press. 1988.

Thompson, Raymond H. 1958. "Preface." In *Migrations in New World Culture History*. Vol Social Science Bulletin No. 27, University of Arizona Bulletin, Volume 29, Number 2, edited by Raymond H. Thompson, pp. v–vii. Tucson, AZ: University of Arizona Press.

Toniello, Ginevra, Dana Lepofsky, Gavia Lertzman-Lepofsky, Anne K. Salomon and Kirsten Rowell. 2019. "11,500 y of human–clam relationships provide long-term context for intertidal management in the Salish Sea, British Columbia." *Proceedings of the National Academy of Sciences* 116 (44): 22106–22114. https://doi.org/10.1073/pnas.1905921116.

Torroni, A., T. G. Schurr, C. Yang, et al. 1992. "Native American mitochondrial DNA analysis indicates that the Amerind and the NaDene populations were founded by two independent migrations." *Genetics* 130: 153–162.

Torroni, Antonio, Rem I. Sukernik, Theodore G. Schurr, Yelena B. Starikovskaya, Margaret F. Cabell, Michael H. Crawford, Anthony G. Comuzzie, and Douglas C. Wallace. 1993. "MtDNA variation of Aboriginal Siberians reveals distinct genetic affinities with Native Americans." *American Journal of Human Genetics* 53: 591–608.

Tosco M. 2000. "Is there an 'Ethiopian language area'?" *Anthropological Linguistics* 42: 329–365.

Townsend, Joan B. 1981. "Tanaina." In *Subarctic, Volume 6 of Handbook of North American Indians*, edited by June Helm, pp. 623–640. Washington, DC: The Smithsonian Institution.

Traill, Lochran W., Corey J. A. Bradshaw, and Barry W. Brook. 2007. "Minimum viable population size: A meta-analysis of 30 years of published estimates." *Biological Conservation* 139: 159–166.

Turner, Nancy J. *Ancient Pathways, Ancestral Knowledge: Ethnobotany and Ecological Wisdom of Indigenous Peoples of Northwestern North America*. McGill-Queen's Native and Northern Series Number 74. Montreal: McGill-Queen's University Press. 2014.

Turner, Nancy J., Chelsey G. Armstrong, and Dana Lepofsky. 2021. "Adopting a root. Documenting ecological and cultural signatures of plant translocations in North America." *American Anthropologist* 124(4): 1–19. https://doi.org/10.1111/aman.13658.

Underwood, Joan V. *Concise English-Korean Dictionary Romanized*. Rutland, VT: Charles E. Tuttle Co. 1954.

Vajda, Edward. 2010. "A Siberian link with Na-Dene Languages." In *The Dene-Yeniseian Connection*, edited by James Kari and Ben A. Potter. Special issue, Anthropological Papers of the University of Alaska 5, pp. 33–99. Fairbanks, AK: University of Alaska.

Volodko, Natalia V., Elena B. Starikovskaya, Ilya O. Mazunin, Nikolai P. Eltsov, Polina V. Naidenko, Douglas C. Wallace, and Rem I. Sukernik. 2008. "Mitochondrial genome diversity in Arctic Siberians, with particular reference to the evolutionary history of Beringia and Pleistocenic peopling of the Americas." *American Journal of Human Genetics* 82: 1084–1100.

Wainman, Newton, and Rolf W. Mathewes. 1987. "Forest history of the last 12,000 years based on plant macrofossil analysis of sediment from Marion Lake, southwestern British Columbia." *Canadian Journal of Botany* 65(11): 2179–2187.

Walker, Carl N. 1973. "An English–Squamish Dictionary based on 'The Squamish Language' Parts I and II, by Aert H. Kuipers." Unpublished Master of Arts Thesis, The University of British Columbia, Vancouver.

Walker, Ian R., and Marlow G. Pellatt. 2003. "Climate change in Coastal British Columbia–A northwestern paleoenvironmental perspective." *Canadian Water Resources Journal / Revue canadienne des ressources hydriques* 28(4): 531–566.

Walker, M. J. C., M. Berkelhammer, S. Björck, L. C. Cwynar, D. A. Fisher, A. J. Long, J. J. Lowe, R. M. Newnham, S. O. Rasmussen, and H. Weiss. 2012. "Formal subdivision of the Holocene Series/Epoch: A discussion paper by a working group of INTIMATE (Integration of Ice-core, Marine and Terrestrial Records) and the Subcommission on Quaternary Stratigraphy (International Commission on Stratigraphy)." *Journal of Quaternary Science* 27(7): 649–659.

Wallace, William, and Donald W. Lathrop. *West Berkeley (CA-ALA-307): A Culturally Stratified Shellmound on the East Shore of San Francisco Bay.* Berkeley, CA: Contributions of the University of California Archaeological Research Facility 29. 1975.

Welch, J. M. An Archaeological Survey of the Chehalis River Valley in Southwestern Washington." Report on file at the Washington State Department of Archaeology and Historic Preservation, Olympia, WA. 1983.

Whistler, Kenneth W. 1977. "Wintun prehistory: An interpretation based on linguistic reconstruction of plant and animal nomenclature." Proceedings of the 3rd Annual Meeting of the Berkeley Linguistic Society, University of California, Berkeley, Berkeley, CA.

Wiberg, Randy 2010. *Archaeological Investigations at CA-CCO-18/548: Final Report for the Vineyards at Marsh Creek Project, Contra Costa County, California.* Manuscript on File, Holman & Associates Archaeological Consultants, San Francisco, CA. 25 September 1977.

Wichmann, Søren, Eric W. Holman, and Cecil H. Brown, eds. 2020. *The ASJP Database (Version 19).* http://asjp.clld.org

Wichmann, Søren, André Müller, and Viveka Velupillai. 2010. "Homelands of the world's language families: A quantitative approach." *Diachronica* 27(2): 247–276.

Wichmann, Søren, and Taraka Rama. 2021. "Testing methods of linguistic homeland detection using synthetic data." *Philosophical Transactions of the Royal Society B* 376: 20200202.

Wigen, R. J. "Vertebrate Fauna." In *The Hoko River Archaeological Site Complex: The Rockshelter (45CA21), 1000–100 B.P., Olympic Peninsula, Washington*, edited by D. R. Croes, pp. 71–104. Pullman, WA: Washington State University Press. 2005.

Willey, Gordon R., Charles Di Peso, William A Ritchie, Irving Rouse, John Howland Rowe, and Donald W. Lathrop. 1956. "An archaeological classification of culture contact situations." *Memoirs of the Society for American Archaeology* 11: 1–30.

Wilson, Andrew M. 2015. "The Aleut kayak: How Aleut technology shaped history." History Undergraduate Theses. 26. https://digitalcommons.tacoma.uw.edu/history_theses/26

Wilson, Darryl, and Lael Wilson. *Suena-English Dictionary and English-Suena Dictionary*. Ukarumpa, PNG: Summer Institute of Linguistics. 2008.

Wohlgemuth, E. 2004. "The course of plant food intensification in Native Central California." PhD Dissertation, Anthropology, University of California, Davis, CA.

Workman, William B. 1998. "Archaeology of the Southern Kenai Peninsula." *Arctic Anthropology* 35: 146–159.

Workman, William B., and Allen P. McCartney. 1998. "Coast to coast: Prehistoric maritime cultures in the North Pacific." *Arctic Anthropology* 35: 361–370.

Zulu Dictionary. Lexilogos. Google Translate. https://translate.google.com// (accessed December 21, 2021).

Index

Page references for entries in figures and tables are *italicized.*
Plants are listed by their common English names, followed by scientific names in parentheses.

About the Authors

Cecil H. Brown is Distinguished Research Professor Emeritus of anthropology and linguistics at Northern Illinois University. His main topical interests include language and culture, comparative linguistics, language typology, and ethnobiology. His principal areal interests are Native Peoples of North, Middle, and South America and their languages, especially Mesoamerican groups. Relevant work: *Lexical Acculturation in Native American Languages* (1999).

Dana Lepofsky is Professor of archaeology at Simon Fraser University. Her main interests are the social and ecological impacts of past human interactions with their environment. She works primarily on the Northwest Coast, in the traditional territories of and in collaboration with several First Nations. Relevant work (Editor): "Indigenous Resource Management: Past, Present, and Future." *Journal of Ethnobiology*, special issue, vol. 29, 2009.

Kent G. Lightfoot is Distinguished Professor Emeritus of anthropology at the University of California, Berkeley. He has undertaken archaeological work in New England, the American Southwest, Alaska, Hawaii, and California. His recent investigations have focused on the archaeology of colonialism, the shell mounds of the greater San Francisco Bay Area, and indigenous landscape and seascape stewardship practices along the central coast of California. Relevant work: Kent Lightfoot and Otis Parrish. *California Indians and Their Environment: An Introduction* (2009).

Nancy J. Turner is Distinguished Professor Emerita in environmental studies at the University of Victoria, British Columbia. She has worked as an ethnobotanist with Indigenous elders and cultural specialists in western Canada for over 50 years, learning about traditional knowledge of plants and environments. She has received prestigious awards for her work, including the Order of Canada, the Order of British Columbia, and fellowship in the Royal Society of Canada, as well as honorary degrees from four BC universities. Relevant work: *Ancient Pathways, Ancestral Knowledge: Ethnobotany and Ecological Wisdom of Indigenous Peoples of Northwestern North America* (2 volumes). McGill-Queen's Native and Northern Series Number 74. (2014).

www.ingramcontent.com/pod-product-compliance
Ingram Content Group UK Ltd.
Pitfield, Milton Keynes, MK11 3LW, UK
UKHW041835031026
15587UKWH00002B/792

9 781666 915105